Pocketguide to

NEW YORK HATCHES

Paul Weamer

HEADWATER BOOKS
STACKPOLE BOOKS

This book is dedicated to my wife Ruthann for everything.

Published by
STACKPOLE BOOKS
5067 Ritter Road
Mechanicsburg, PA 17055
www.stackpolebooks.com

Distributed by NATIONAL BOOK NETWORK

Library of Congress Cataloging-in-Publication Data

Weamer, Paul.

Pocketguide to New York hatches / Paul Weamer.

pages cm
Includes bibliographical references and index.

ISBN 978-0-8117-3170-6 (hardcover)
1. Flies, Artificial—New York (State) 2. Aquatic insects—New York (State)—Identification. 3. Fly fishing—New York (State) 4. Trout—Food—New York (State) I. Title.

SH451.W43 2013
799.12'4097471—dc23

2012047781

Contents

Acknowledgments

Thanks to James McClure, another bug nerd like me, for collecting several of the aquatic insect species included in this book and for not being upset when I shot bug photos rather than fished during our outing.

Thanks to Patrick Williams for collecting the *Dolophilodes* caddis.

Thanks to Jake Villwock for the foam Golden Stonefly patterns and to Brian Wolfkiel, Rob Lewis, and Lee Hartman for your stonefly nymph patterns.

Thanks to Ken for generously supplying me with petri dishes and collection vials, which were a tremendous aid during my work on this book.

Thanks to Dan "Guru" Morris, Mutton Hollow Outfitters, and all of the other customers of the fly shops I've owned and managed, for your support and encouragement.

Thanks to Jay Nichols and Judith Schnell for allowing me the opportunity to write another Stackpole/Headwater book. And thanks to my terrific editor Amy Lerner and to Trish Manny for your help.

Thanks to Ross Purnell and John Randolph for giving me the opportunity to write about aquatic insect hatches for *Fly Fisherman*.

Thanks to Orvis, Regal, and Simms for your support and for making the finest fly-fishing and fly-tying products in the world.

Thanks to my father, Ed Weamer, for helping me collect and shoot photographs of aquatic insects on my fortieth birthday, rather than actually fish.

Thanks to Fran Betters. Though Fran has now passed on, I will never forget his kindness and willingness to help me when I started in the fly-fishing industry. Fran Betters was one of America's fly-fishing and fly-tying giants, and he will never be replaced.

Thanks to Mary Dette Clark and Joe Fox of the Dette fly shop for your years of kindness and friendship and for allowing me to drink your coffee, track mud onto your carpet, and shoot photographs of your many beautifully tied flies.

Thanks to Matt and Sam of West Branch Anglers and Joe and Coz of Border Water Outfitters for your help.

Thanks to my good friend and Penn State University entomologist Greg Hoover, perhaps the most knowledgeable angling entomologist in the United States, for your help collecting *L. recurvata* larvae and *E. simulans* larvae and duns, for the rainy-day trip to New York, dodging the ghost coyote, and for years of answering my many questions about aquatic entomology. And thank you Greg for providing me with your amazing photographs of the Trico and *Caenis* larva, *P. adoptiva* larva, *Hexagenia limbata* dun, *Capniid* larva, and *Strophopteryx* larva . . . notice I wrote larva, not nymph. I knew you'd like that.

Thanks to my very good friend, Charlie Meck, for allowing me to include your phenological insights in this book, for cowriting this book's predecessor with me (*Pocketguide to Pennsylvania Hatches*), and for the many years of teaching me patiently much of what I know about aquatic entomology.

—Paul Weamer

Introduction

Many people who have read my other books have questioned me about this project. Charles Meck and I wrote a book called the *Pocketguide to Pennsylvania Hatches* in 2009. So why write one for New York? After all, the insects are the same and so are many of the fishing techniques. What is the purpose of a pocketguide dedicated solely to New York's hatches?

I have several reasons for writing this book. First, in the Pennsylvania pocketguide, Charlie and I decided to include photographs of only adult insects. We believed (I still believe) that the average angler has a much easier time identifying an insect if he's looking at an adult specimen. Many mayfly and stonefly nymphs, as well as caddis larvae, look very similar from one species to the next, and some can be identified to the species level only with a microscope. Most people have neither the time nor the inclination to do that. In spite of this, including nymph and larvae photographs was one of the most common requests from readers of the PA pocketguide.

So in this edition, I add photos of nymphs and larvae as well as the mayfly adult form, commonly called a spinner, caddis cases (for those that make them), and undershots of adult caddis bodies. In addition, the insect entries are expanded to include several mayflies, caddisflies, stoneflies, and other aquatic insects that weren't included in the Pennsylvania book.

You'll find detailed information for each important New York trout food: where a particular insect will appear and what time of day you can expect to find it, along with specific information about what the insect looks like and how best to catch trout by imitating it.

In addition to describing the most important mayflies, caddisflies, and stoneflies, I highlight other aquatic insects and crustaceans that trout sometimes eat: midges, crane flies, scuds, sow bugs, and others. Though fly fishermen don't regard these insects as highly as mayflies, stoneflies, and caddisflies, they are sometimes just as, or even more, important for catching fish.

Finally, I discuss the most common and important terrestrials along New York's trout streams. There are times when these creatures can also be extremely important for catching fish in New York's streams and rivers.

If you read the *Pocketguide to Pennsylvania Hatches*, you'll notice that the fly pattern recommendations here have been altered. When it comes to match-the-hatch dry-fly fishing, New York has led the nation in developing innovative fly patterns. While it is true that the flies included in the Pennsylvania book would also work well in New York, I believe that flies developed in New York, specific to New York waters, add a depth that you wouldn't find if you were using the Pennsylvania guide to fish the Catskills' Beaverkill or the Adirondacks' Ausable.

I do not list a specific spinner, nymph, or larva fly pattern for every insect. To do so would make this book so large that it would have been cumbersome to carry streamside. But that's not the only reason. Nymphs or larvae of varying mayfly, stonefly, and caddis species often share the same basic form, and most mayfly spinners vary only in color and size. Usually, a basic spinner imitation or general nymph pattern such as a Pheasant Tail, Hare's Ear, or standard bead-head larva, tied with materials dyed to match the naturals' color, can be used to imitate any of them. But I describe them in depth and include many photos of caddis larvae, stonefly nymphs, and mayfly spinners and nymphs so that you can adapt my basic nymph and spinner selection to match any of these insects.

1 Using the Hatch Charts

You can find hatch charts in every fly-fishing entomology book and many fly-fishing websites. But hatch charts are static, and they cannot factor in the dramatic effects of the weather on aquatic insect emergences, so in many years, the timing for hatches in the charts is wrong. For example, in the last two years, upper Penns Creek has had Green Drake hatches beginning on May 17 in 2012, when it was unseasonably warm and dry, and in the first week of June in 2011, when it was unseasonably cold and wetter than normal. That's a three-week variation for one species in two consecutive years, and there isn't a hatch chart in existence that could have predicted it.

Another problem with hatch charts is that their information is too general. When they list an approximate date for an insect emergence, what does that date actually mean? For instance, are the hatch charts intended only for primary, famous rivers and streams like the lower Beaverkill? What about the upper Beaverkill or its tributaries? About ten years ago, when I was fishing the upper Beaverkill at the DEC campground, I encountered an unexpected Green Drake emergence. The hatch had already ended about two weeks earlier on the lower Beaverkill. This past spring, I found a Little Brown Stonefly adult flying along a small tributary stream in early May, well after the emergence date listed in most hatch charts.

There's a reason for this hatch timing discrepancy. Headwaters of big, famous trout streams and rivers are much colder than their lower sections. So are many of their small, tree-lined, spring-fed tributaries. It takes these sections much longer to warm to proper emergence temperatures for their aquatic insects. And you can't

Ruthann Weamer with a beautiful New York Brown trout taken during a heavy mayfly hatch in the Catskill region.

pack all of that information into one hatch chart.

Geography also plays a part in hatch timing. The geographic regions of the Catskills, the Adirondacks, and western New York are too vast for each aquatic species in each body of water to hatch simultaneously. Generally speaking, hatches on streams in northern regions, or at a higher elevation, commence later in the year than hatches on streams in southern regions or at lower elevations. So hatches on trout streams in the southern Adirondacks happen earlier than those in the northern Adirondacks.

This book contains hatch charts for each of the major aquatic insect families—mayflies, caddisflies, and stoneflies—with separate emergence weeks for each of the three regions of the state: western New York, the Catskill region, and the Adirondack region. Remember that the dates vary for every river and stream within these regions, so it's a good idea to keep notes on the streams you fish and develop your own charts for each watershed.

Western New York, west of the Adirondack and Catskill mountains, has some great, but unheralded, fly-fishing opportunities. Because it's near the Great Lakes and is lower in elevation, it has a a warmer climate, and hatches often begin there before they do in the two mountainous regions. In fact, the climate is so mild that grapes grow on the shores of the Great Lakes, although winter snows, caused by cold air traveling over warm Great Lakes water, can be severe, dumping some of the largest snowfall totals in the contiguous United States each year.

The Catskill region is vast, but not as warm as western New York or as far north and, therefore, as cold as the Adirondack region.

The Upper Delaware's West Branch is one of the most important trout fisheries in the Catskill region. RUTHANN WEAMER PHOTO

Spring comes earlier to the Catskills than the Adirondacks. The Catskill mountain peaks are not as high, but 98 of them rise over 3,000 feet. The Catskill Park is a massive area that is 60 percent privately owned and includes many small hamlets and villages. The Catskill Forest Preserve section of the park was formed in 1885 and is entirely public–its "Forever Wild" forests encompass nearly 300,000 acres.

The Catskills are one of epicenters of American fly fishing. The Beaverkill, Willowemoc, Neversink, Esopus, Schoharie, East and West Branches of the Delaware, and many other rivers and streams celebrated in classic American fly-fishing literature flow through the Catskills. Today its rivers and streams still offer some of the best fly fishing east of the Mississippi River.

The Adirondacks feel more like Maine than New York. The Adirondack Park, created in 1892, is a vast eastern wilderness made up of 6 million acres, a combination of 2.6 million acres of "Forever

Wild" state-owned land that can never be developed and 3.4 million acres of private property. The park is bigger than Glacier, the Grand Canyon, the Great Smoky Mountains National Park, Yellowstone, and Yosemite combined. One million of the Adirondack Park's acres are classified as wilderness, making it the largest state-controlled wilderness area in the lower 48 states.

This is eastern high country, and it boasts the highest mountain in New York state, Mount Marcy, which rises 5,343 feet into the heavens. The Adirondack Park has 3,000 lakes and ponds and 30,000 miles of rivers and streams which, combined, offer more spots than one angler can hope to fish in a lifetime. This amazing resource of trout-holding waterways has its own history, every bit as important as the Catskills'. But accurate emergence dates for every hatch on every pond, lake, river, and stream in the Adirondacks is impossible because the area is so vast.

Perhaps the most important information in this book's hatch charts is the order in which the hatches occur. Though it's impossible to list when the Hendricksons will begin emerging in your favorite New York water, I can say with certainty that they will hatch before the Green Drakes will—and that may be the best way to use the charts. Study the insects in your waters, contact local fly shops if you can, and follow the hatch progression within the charts.

Phenology, which studies the relationship between plants and animals (including insects), is a much more reliable hatch indicator than a date in a chart. My good friend Charlie Meck has graciously allowed me to use his phenology indicators for this book. I cannot tell you when the Hendricksons will hatch in your region of New York next year, but I do know that you will begin to see them on the water when the forsythia is in bloom. If you look at the plant indicators in this book, you won't have to wonder when a hatch will begin. The plants will let you know.

Though the hatch charts list a few mayfly, caddisfly, and stonefly species of minor importance in New York, I have not included in-depth discussions or photographs for most of these insects. It would be impossible to include all of the potential aquatic and land-born insects that a trout might eat in New York state and still

The Adirondacks' Ausable River is legendary for trout fishermen. Thankfully much of it is easier to wade than the turbulent Ausable Chasm.

keep the book small enough that you can easily carry it while you're fishing. And that is the purpose of this book, so I spent a great deal of time pondering which insects to include. Though I realize that I may have omitted a few insects that may be locally important, I'm confident that I have thoroughly covered the important insects throughout the state.

2 Using This Guide

ABOUT THE ENTRIES

Description. Colors, number of tails, and other major identifying features, including any differences between male and female. This information provides the basis for both identification and replication at the tying bench. Includes notes on any other insects that this one resembles.

Emergence. A description of when the insect hatches, including flowers and trees that are blooming and other hatches that might appear at the same time. The dates given are just estimates and are subject to variables in the weather, water temperature, and type of stream, among other things. Time of day is an average or description of what is typical. When it is cloudy and cool, Hendricksons can hatch all day on some streams; on a bright sunny day, the hatch may last only a short period.

Behavior and Tactics. Any behavior that is important to anglers and tactics for imitating that behavior. I often suggest flies and how to fish them, but you should refer to the following section on aquatic insect and fly pattern color and match your fly selection based on the conditions and your observations of the naturals.

Notes. Additional information of interest to anglers about the insect or the accompanying photos.

---

1. The most common name of the insect in New York.
2. Other common names for the bug.
3. Genus and species, followed by previous (prev.) names.
4. Approximate week the hatch begins in each of New York's three regions, how long the hatch lasts, and the time of day it occurs.
5. Approximate hook size range. Assume standard-shank hooks unless otherwise noted. Approximate millimeter range is included because of the variance among hook models. The size of bugs can vary from stream to stream and even on different sections of the same stream.

AQUATIC INSECT AND FLY PATTERN COLOR

Some anglers and fly tiers are obsessed with imitating exact mayfly coloration. They never find the proper dubbing color for Sulphurs (or any other hatch) on a fly shop peg-board, so they meticulously mix bits of colored dubbings, pseudo-scientifically creating their own perfect color match for each specific hatch. If these anglers fail to catch fish during a Sulphur emergence, they don't believe that it was their technique, their fly pattern choice or size, or their tippet size or length that failed them; they know it was their fly's color.

But fly fishing and tying are not exact sciences. Though much of our sport has its roots in natural science, fly fishing will always have an X factor, which cannot be explained. Why can two anglers, fishing a hatch together, have similar results even if they use different flies tied with different materials? If color was the determining factor in trout selectivity, then wouldn't both anglers have to be using identically colored flies to catch fish?

The author with a New York brown trout caught during an intense Sulphur hatch.

Mayflies often have significant color variances, even among the same species in the same creek. Slate Drakes are a good example. Some Slate Drakes are gray with a hint of olive or maroon on their undersides. Some Slate Drakes are a reddish mahogany color, while yet others are bright green or olive when they first emerge. Most mayflies experience slight color changes after emerging, darkening as they are exposed to the air. Some undergo dramatic changes in color. So mayflies that are hatching right in front of you are often lighter, or paler, than those that have hatched upstream and floated down to you.

Most of New York's public stream accesses are marked with signs like this one.

Some fly fishers take fly color to the opposite extreme. These anglers believe that the only fly needed to catch trout is a gray-bodied Adams (or any other basic attractor pattern). The Adams is a great fly pattern that has been responsible for many fly-caught trout. And on a sterile freestone stream, or a remote waterway where the fish see few fly patterns, it may be all that you need to catch trout. But on New York's highly pressured rivers, creeks, and tailwaters, a gray Adams might not work when you're imitating light-colored Sulphurs. Sure, the occasional fish might see an Adams as a floating nymph and eat the fly. But fish that are selectively feeding to light-orange duns and emergers in pressured waters often ignore patterns that are darkly colored.

People who identify bugs for a living might have the best answer for the fly color debate. Professional entomologists—real bug experts—place very little emphasis on a fly's color when determining its species. Color is unimportant to these scientists because it is too variable, even among the same species, to provide positive identification.

Water chemistry, food choice and options during the nymph stage, and minute genetic differences can all affect aquatic insect color. Just look at your own family. Does your father, mother, or sibling have a slightly different skin tone from your own? Why wouldn't insects have these same differences? Any of these factors can collide in a stream to create subtle variances in insect coloration, making it impossible for anglers to imitate each mayfly's exact color.

Nature doesn't produce clones. It creates a pantheon of shades for its living organisms to wear. The underside of most Green

Jim Costolnick and Geoff Mueller with a nice Delaware River brown trout.

Drakes is cream with a hint of yellowish olive, so you wouldn't want to imitate Green Drakes with dark orange dubbing. But if you don't scare the fish by wading too loudly, and you throw a good cast with the proper mend so that your fly floats over the fish without dragging, then just plain old cream will probably work too.

Use the fly pattern recipes in this book as guidelines for tying your own flies. Concentrate your efforts on imitating the shades of color for the mayflies in your favorite stream rather than the exact color—which doesn't exist anyway—and your flies will do their part to catch fish.

3 Mayflies

Scientists group mayflies in the order Ephemeroptera, a name that comes from "ephemeral" or "short-lived," because most mayflies live only a few days out of the water. Most mayflies spend approximately 362 days underwater as eggs and nymphs but barely three days as duns and spinners. When the female spinner lays the fertilized eggs, it takes several weeks for the larvae (called nymphs) to hatch. Eggs of some species like Tricos and White Flies that are laid in the fall delay emerging from an egg to a nymph until spring so the young do not have to endure a long, cold winter. Once hatched, nymphs search out and live in specific areas of a stream. Species like the Quill Gordon and March Brown live under rocks; others, like Slate Drakes, live in open view on the bottom of the stream. Many members of the family Ephemerellidae live in aquatic weeds, while Green, Yellow, and Brown Drakes and White Flies burrow into the substrate or bottom of the stream. Most nymphs feed and grow for about a year before they change into air-breathing duns. Nymphs grow by splitting and shedding their hardened pellicle or nymphal skin in a series of stages called instars.

While most mayflies live underwater for about a year, some spend a shorter amount of time under the water. Tricos, for example, have two generations each year. The eggs of this and similar species, when laid in late October, don't hatch into larvae until May. Once the nymphs emerge, they live underwater for about seven weeks (48 days).

Some mayfly nymphs live underwater more than a year. Green Drake nymphs live underwater for two years, so the progeny of Green Drakes emerging in 2013 will emerge in 2015. Many of the

Epeorus vitreus nymph

Mayfly and stonefly nymphs often appear white for a brief time after molting (an instar). Some anglers use white nymph patterns to imitate this trait.

larger stoneflies also take two years to emerge as adults, and some even three. This is why you'll sometimes see especially heavy hatches one year and much lighter ones the next. Conditions like droughts or floods can adversely affect the flies of one year while doing little harm to the following year's crop.

Some mayflies transform from nymphs to air-breathing duns near the stream bottom, a few in the middle, and others at or near the surface; still others crawl onto rocks or streamside vegetation first. Where they change can be critical for the angler. If the nymph changes near the bottom—see the Quill Gordon (*Epeorus pleuralis*) and the Pink Lady (*Epeorus vitreus*)—then fishing wet flies to imitate the dun rising to the surface is a dynamite way to catch trout.

Other nymphs, like the Slate Drake (*Isonychia bicolor*), swim to a rock near the shore, and Gray Drakes (*Siphlonurus* sp.) swim to streamside vegetation and crawl out of the water to emerge. So moving the nymph toward the shore often works when you fish these hatches.

The emerging nymph, or emerger as anglers call it, is one of the two most vulnerable stages of the mayfly life cycle (the other is the spent spinner). Once the emerger sheds its pellicle and becomes an air-breathing subadult, it is called a dun or, as scientists call it, a subimago. The subimago is usually not capable of mating.

When mayfly duns emerge, they rest on the surface until their wings fill with fluid, allowing them to fly. When the water and air are colder, they ride on the surface longer because it takes longer for the fluid to fill their wings. If duns are greeted with cool weather and a bit of a drizzle, it's even better for the fly fisher, since both conditions slow the mayfly down considerably. On these cool, drizzly days, mayflies ride on the surface for an extended period, providing trout with a greater opportunity to feed on them.

After the dun takes flight, it heads toward a tree or bush to rest. The mayfly dun usually rests on a leaf or branch for at least an hour to two and sometimes for up to three days, depending on the species and the weather. On cool spring days, for example, it might take a couple days until the dun changes to a spinner. Some species, like the Dark Quill Gordon (*Ameletus ludens*), don't fly to trees but skitter across the surface and rest on rocks or other debris near the shore. The wings of this particular species are poorly developed because the species is parthenogenic: In other words, the female does not need to mate to lay fertilized eggs, so this species has no mating flight.

Finally the dun changes to a mating adult, which scientists call the imago and anglers call the spinner. A very few species like the female White Fly (*Ephoron leukon*) don't change from dun to spinner, but mate as duns. The male White Fly, however, does change to a spinner. Changing from a dun to a spinner might take the White Fly and Trico a few minutes, but it can take many other species a day or more.

The male and female spinners appear in swarms over faster sections of streams or near trees where they rested. They mate while still in flight, the male holding on to the female with his clasper (two pincer-like appendages beneath his tail). Mated female adults of some of the larger mayflies, like the Brown and Green Drake, move upstream in huge formations 10 to 20 feet above the surface.

Male Tricos swarm over faster sections of a stream, and the females enter the swarm and mate and then move upstream to lay their eggs. This is one way mayfly species ensure that the entire stream will hold future populations: Upstream eggs move downstream in the water's current, so if females flew downstream to lay their eggs, over the generations the eggs would be washed into the ocean and the hatch would disappear.

Another way that mayflies ensure their survival is the order in which males and females hatch. For most mayflies, males of the species are the first to hatch, and one male can fertilize more than one female. If females hatched first, and a catastrophic weather event such as a severe frost killed them all, the males would hatch without mates. But if a cataclysmic weather event occurs after the male hatch, but before the female hatch, the consequences are much less severe. As long as a few males remain, they can fertilize all of the females and the species can remain viable.

Females lay their eggs in several ways. Some species, like Little Blue-Winged Olive duns, actually dive under the surface and scatter their eggs. Other female spinners fall onto the surface and lay their eggs; still others drop their eggs while in flight. The more you understand the way each species lays its eggs, the better prepared you can be to fish the spinnerfall.

The best matching-the-hatch events occur with species that drop to the surface and lay their eggs. Spinners die as they dive or sink under the surface, making sunking spinner fly patterns important for imitating the naturals.

The eggs hatch in a few weeks to a few months, and the mayfly life cycle is completed.

The mayfly and stonefly life cycles are incomplete life cycles because they lack a fourth stage, the pupa or "resting" stage. Cadisfly (Tricoptera) life cycles include this resting stage.

MAYFLY LIFE CYCLE

MATURE AND IMMATURE NYMPHS

Immature Green Drake nymph

Mature Green Drake nymph

General Mayfly Nymphs

Hare's Ear, BDHD Pheasant Tail, Copper John, BDHD Prince Nymph (See the appendix for recipes.)

General Mayfly Spinner Patterns

Poly Wing Spinner, Grizzly Hackle Wing Spinner, Marinaro Cream Hackle Wing Spinner (tied by Vince Marinaro), CDC Wing Spinner (See the appendix for recipes.)

Hatch-Matching Mayfly Nymphs

Humphries Sulphur, Williams Dark Mayfly Nymph, Drake Nymph, Stroup's Slate Drake, March Brown (See the appendix for recipes.)

MAYFLY HATCH CHART

Scientific name	Common name/size	Time of day	Hatch begins Western regi
Baetis tricaudatus	Little Blue-Winged Olive #18-22	afternoon	second week March, third week of September
Paraleptophlebia adoptiva	Spring Blue Quill #16-18	early afternoon	first week of April
Epeorus pleuralis	Quill Gordon #12-14	early afternoon	first week of April
*Ameletus ludens**	Dark Quill Gordon #12-14	early afternoon	second week April
*Siphloplecton basale**	Dark Olive Dun #12-14	afternoon	second week April
Ephemerella subvaria	Hendrickson #14-16	afternoon	second week April
Leptophlebia cupida	Black Quill #12-14	afternoon	fourth week o April
Acentrella species*	Little Blue Dun #20-24	afternoon	first week of May (first brood), Augus through Octo (later broods)
Ephemerella invaria	Sulphur #14-16	evening	late in first we of May
Maccaffertium vicarium	March Brown #10-14	afternoon and early evening	second week c May
*Leucrocuta aphrodite**	Pale Evening Dun #16	early evening	second week o May
*Ephemerella septentrionalis**	Pale Evening Dun #16	early evening	second week o May

*Minor hatches

Hatch begins in Catskill region	Hatch begins in Adirondack region	Hatch duration
rth week of rch, fourth week September	third week of April, second week of September	one month (first generation), two weeks (second generation)
ond week of April	first week of May	two weeks
ond week of April	late in first week of May	ten days
rd week of April	second week of May	ten days
rd week of April	second week of May	ten days
rd week of April	second week of May	two weeks
st week of May	third week of May	one month
cond week of May irst brood), August rough October ater broods)	fourth week of May (first brood), August through October (later broods)	three weeks (first brood), two months (later broods)
cond week of May	late in third week of May	three weeks
ird week of May	late in third week of May	six weeks
ird week of May	first week of June	two weeks
ird week of May	first week of June	two weeks

MAYFLY HATCH CHART *continued*

Scientific name	Common name/size	Time of day	Hatch begins Western regi
*Eurylophella bicolor**	Chocolate Dun #16	afternoon	third week of May
Ephemerella dorothea dorothea	Little Sulphur #18-20	evening	third week of May
Ephemera guttulata	Green Drake #8-12	evening	third week of May
Litobrancha recurvata	Dark Green Drake #6-8	afternoon	third week of May
Isonychia bicolor	Slate Drake #10-14	early evening (first generation), afternoon (second generation)	third week of May, second week of September
Ephemera simulans	Brown Drake #10-12	evening	fourth week o May
Drunella cornuta, D. cornutella, and *D. lata*	Blue-Winged Olive #14-16 (*D. cornuta* and *D. cornutella*); #16-18 (*D. lata*)	late morning and early afternoon	first week of June (later for *D. lata*)
Siphlonurus species	Gray Drake #12-14	afternoon	first week of June
Stenacron interpunctatum	Light Cahill #14	early evening	first week of June
*Teloganopsis deficiens**	Dark Blue Quill #18-20	early evening	first week of June
*Ephemerella needhami**	Olive Sulphur #16	evening	first week of June

*Minor hatches

Hatch begins in Catskill region	Hatch begins in Adirondack region	Hatch duration
›urth week of May	first week of June	two weeks
›urth week of May	first week of June	three weeks (longer in tailwaters)
›urth week of May	late in first week of June	one week
›urth week of May	second week of June	approximately one week
›urth week of May, ›econd week of ›eptember	second week of June, third week of August	one month (first generation), two weeks (second generation)
›irst week of June	second week of June	less than one week
›econd week of June later for *D. lata*)	third week of June (later for *D. lata*)	ten days (several weeks for *D. lata*)
›econd week of June	second week of June	two weeks
›econd week of June	second week of June	one month
›econd week of June	third week of June	two weeks
›econd week of June	third week of June	one week

MAYFLY HATCH CHART *continued*

Scientific name	Common name/size	Time of day	Hatch begins Western regic
Maccaffertium and *Stenonema* species	Cahill (cream and dark) #14-16	early evening	first week of June
Ephemera varia	Yellow Drake #10-12	evening	second week o June
Epeorus vitreus	Pink Lady (Pink Cahill) #12-16	evening	second week o June, first wee of September (second generation)
Paraleptophlebia species	Summer Blue Quill #18-20	morning	second week of June
Hexagenia limbata	Hex #6-8	evening	third week of June
Anthopotamus distinctus	Golden Drake #10-12	evening	late in third week of June
Tricorythodes species	Trico #22-26	morning	late in first wee of July
Hexagenia rigida	Hex #8	evening	second week of July
Leucrocuta hebe	Pale Evening Dun #16-18	afternoon and evening	first week of August
Ephoron leukon	White Fly #12-16	evening	second week of August
Hexagenia atrocaudata	Big Slate Drake #8-10	dusk	late in second week of August

*Minor hatches

Hatch begins in Catskill region	Hatch begins in Adirondack region	Hatch duration
econd week of June	third week of June	three months
hird week of June	fourth week of June	six weeks (but most duns hatch in 10 days)
hird week of June, irst week of eptember (second eneration)	fourth week of June, fourth week of August (second generation)	two weeks (first generation), one week (second generation)
hird week of June	fourth week of June	three months
ate in third week of une	late in fourth week of June	two weeks
ourth week of June	second week of July	ten days
econd week of July	second week of July	twelve weeks
hird week of July	third week July	two weeks
irst week of August	second week of August	ten weeks
second week of August	third week of August	one week
late in second week of August	late in third week of August	one week

Little Blue-Winged Olive

Male Little Blue-Winged Olive dun

RS2

page 188

aka *Baetis*, BWO, Blue Dun
Baetis tricaudatus, prev. *B. vagans*
Western NY: second week of March, third week of September
Catskills: fourth week of March, fourth week of September
Adirondacks: third week of April, second week of September
Duration: one month (first generation), two weeks (second generation)
afternoon
#18–22 (5–8 mm)

Description

Nymphs: (Swimming) Tan tails (3), light olive-tan body and legs with dark brown markings, dark brown wing pads. **Dun:** Gray tails (2) with olive reflections, olive-brown belly and legs, slate backs, and medium slate-gray wings. **Spinner:** Dark grayish-brown tails (2), dark rusty brown thorax and legs, rusty tan abdomen with

Male Little Blue-Winged Olive nymph

Male Little Blue-Winged Olive spinner

lighter ribbing, and clear wings. Resembles many other *Baetis, Plauditus,* and *Acentrella* species.

Emergence

Sporadic hatches may begin as early as late February, but March and April emergences are more common; emerges when forsythia is heavily budded and coltsfoot is just starting to bloom. A second generation of smaller *Baetis* hatches in September and October when fall aster is blooming. In colder weather, *Baetis* emerge in the late morning to midafternoon and during warmer weather they emerge in the morning or near dark.

Behavior and Tactics

Early in the season, dead-drift tannish-olive Hare's Ear Nymphs near vegetation and through riffles. Closer to and during the hatch, the nymphs darken and are better matched by dead-drifting Pheasant Tails. *Baetis* nymphs drift or swim in the current until they reach the water's surface. During cold weather, RS2s, Snowshoe Bucktail Bodies, and CDC Burke Emergers imitate *Baetis* struggling out of their nymphal shucks, trying to take flight. To match the duns, use low-riding dry flies such as parachutes or CDC Comparaduns. If the air temperatures are warmer than usual, the duns are more active and hackled patterns do a better job of creating the illusion of increased movement. If you are not sure what stage the trout are feeding on, attach a #18-20 Pheasant Tail to your dry fly to give the fish options. Use soft-hackles and sunken spinners to imitate the *Baetis* spinners drifting helplessly in the current after they dive under the surface to lay eggs.

Notes

Baetis are prolific in weedy, fertile waters, but they are one of the most common insects in the state because they are also able to colonize a wide range of habitats. This mayfly makes up for its small size with its numbers. *Baetis* sometimes emerge simultaneously with Spring Blue Quills, Quill Gordons, and Hendricksons, but their emergence is particularly tied to Spring Blue Quills. Little Blue-Winged Olives often hatch first, then Spring Blue Quills hatch in equal numbers for an hour or two, until the *Baetis* hatch wanes and only the Spring Blue Quills remain. On days when the water temperature is above 50 degrees but the air temperatures are significantly colder, *Baetis* wings often take a long time to be ready for flight, keeping the duns on the surface and providing superb fishing.

The *Baetis* nymph in the photo is a male: notice the large, red eyes. Female nymphs have much smaller, black eyes. It's not always that easy to tell male from female nymphs with other mayfly species.

Baetis look very similar to the tiny *Acentrella* Blue-Winged Olives (formerly *Pseudocloeon* or Pseudos). But you can easily tell the two apart. *Baetis* nymphs have three tails, *Acentrella*, two. *Baetis* duns and spinners have a small hind wing and *Acentrella* do not have a hind wing. But luckily for anglers, fishing methods and tactics for *Baetis* also work for *Acentrella*.

Spring Blue Quill

Male Spring Blue Quill dun

aka Paralep, Little Mahogany Dun
Paraleptophlebia adoptiva
Western NY: first week of April
Catskills: second week of April
Adirondacks: first week of May
Duration: two weeks
early afternoon
#16–18 (6–8 mm)

Conover (Mary Dette), page 188

Description

Nymph: (Swimming) Tails (3) are light brown near body and tan toward the tips; dark reddish-brown body with tan legs, and dark gray to black wing pads. **Dun:** Dark grayish-brown to medium gray tails (3), mahogany to dark brownish-gray body and legs, and slate-gray wings. **Spinner:** Gray tails (3), dark reddish-brown body and legs, clear wings. Resembles *Ephemerella deficiens.*

PHOTO BY GREG HOOVER

Above: Spring Blue Quill nymph; *Above right:* Female Spring Blue Quill dun; *Right:* Male Spring Blue Quill spinner

Emergence

This small mayfly appears in large numbers near the beginning of the season, when forsythia is heavily budded with some flowers already open and trout lilies are blooming. It is often a concentrated hatch, lasting only a couple hours from late morning until early afternoon. But if it's unusually warm, hatches may continue until evening. Once it begins, expect daily hatches for a week or more.

Behavior and Tactics

Early in the season, olive brown Hare's Ears dead-drifted through riffles imitate nymphs that are immature (not yet ready to emerge). Mature nymphs (those that are ready to emerge) are much darker, and Pheasant Tails fished along the edges between fast and slow-moving water, with a dead drift or slow stripping retrieve toward shore, imitate naturals that migrate from riffles to slower water to emerge. Emergers swim to the surface to shed their nymphal shucks in, or near, the film. Cooler, inclement weather, common in early April, slows the newly hatched duns' escape from the surface, and emergers tied with CDC and snowshoe rabbit are effective at this time. As with *Baetis,* use flush-floating dry flies to imitate placidly floating duns, and patterns with hackle to imitate duns that are moving more (typical in warmer weather). Spinners fall in the late afternoon or early evening.

Notes

P. adoptiva often emerges with Little Blue-Winged Olives, Hendricksons, Quill Gordons, and Grannoms. Unlike summertime Paralep species, the *adoptiva* male spinner looks similar to the female. In Summer Blue Quill species the male spinner's abdomen becomes clear (it looks similar to a Trico) and anglers often call it the Jenny Spinner, but the female remains reddish brown.

Though the Conover is most often used today to imitate Hendricksons, and was originally tied to imitate caddisflies, it provides an excellent hatch-matching choice for Spring Blue Quills when tied in smaller sizes.

Notice that the female duns are much more mahogany in color than the darker males, inspiring the angler's name Little Mahogany Duns.

Spring Blue Quill and Summer Blue Quill nymphs look very similar, though Spring Blue Quills' backs are a little lighter. If you tie flies to match the Summer Blue Quill nymph (page 78), and fish them in the spring, the fish won't know the difference.

Quill Gordon

Quill Gordon Emerger

page 188

Epeorus pleuralis, prev. *Iron fraudator, pleuralis*
Western NY: first week of April
Catskills: second week of April
Adirondacks: late in first week of May
Duration: ten days
early afternoon
#12–14 (9–12 mm)

Description

Nymph: (Clinger) Dark olive-brown tails (2) with amber banding, dark olive-brown body, and legs that have areas of amber and heart-shaped markings on the femurs. **Dun:** Dark gray tails (2); dark gray to tannish-olive-gray bodies with olive reflections (body has lighter gray ribbing, and legs have darker heart-shaped markings on the femurs); and slate-gray wings. **Spinner:** Dark brown to black tails (2), reddish- to tannish-brown body with tannish-cream rib, reddish-brown legs, and clear wings with prominent black venations. Resembles *Ameletus ludens* and *Epeorus vitreus* (male dun only).

Emergence

These first large mayflies of the season usually begin hatching when the forsythia are just about ready to open. Hatches usually commence at the most comfortable time of the day, around 1 or 2 PM, and generally continue for one or two hours.

Behavior and Tactics

Heavily weighted, flat, dark amber to brown Hare's Ears tied to imitate the wide, flat bodies of the naturals and fished with a dead drift through riffles can be effective before the hatch begins. Poor swimmers, the nymphs crawl from fast currents into slow or slack water to emerge, and the same heavily weighted nymph patterns slowly stripped from faster current toward slack water can be effective at this stage. The nymphs usually hold tight to submerged rocks and emerge on the stream bottom, rising to the surface as fully formed duns. To match this, swing two wets or a wet fly and a nymph from fast water into slack water areas. You can weight the flies or use split shot on your leader.

After the duns emerge, their substantial wings often take a long time to be ready for flight, because of their subsurface emergence and the cold, raw weather common in April. Duns often stand erect on their legs, a posture best imitated by flies with stiff hackle. Blind casting dry flies is seldom effective for this hatch, so stalk rising fish. Most of the time the spinnerfalls are sparse (commonly occurring

Quill Gordon nymph

Quill Gordon nymph

from 12 to 2 PM) and do not inspire as much feeding activity as the nymphs or duns, though at times fish do key on them.

Notes

Epeorus pleuralis received its common angling name "Quill Gordon" because it looks very similar to the Gordon Quill—a dry-fly pattern designed by Theodore Gordon. Mr. Gordon was a Pennsylvania native, born in Pittsburgh, but he became a fly-fishing legend in New York's Catskill Mountain region early in the twentieth century because of his writing and his work creating and refining the Catskill-style dry fly.

Though Quill Gordons hatch on the stream bottom and swim to the surface without their nymphal shuck, it can still be important to include the shuck in fly patterns. This is a triggering mechanism for trout, indicating an easy meal. But if your shuck yielding patterns are rejected by more discriminating fish, then try a pattern without one.

Not only are male Quill Gordon eyes much larger than the female's, but the color varies. The eyes of many mayfly species change color from black and brown to powdery shades of gray, blue, and green when they are exposed to light. Quill Gordon spinners look very similar to Hendrickson spinners, and they are often found together over riffles in the spring. But you can tell the two species apart easily. Quill Gordon spinners have only two tails and Hendricksons have three.

Intolerant of water pollution, Quill Gordons are found only in the cleanest waters.

Hendrickson

Male Hendrickson dun

DK Hendo Emerger

page 189

aka Red Quill, Light or Dark Hendrickson
Ephemerella subvaria
Western NY: second week of April
Catskills: third week of April
Adirondacks: second week of May
Duration: ten days
afternoon
#14–16 (9–12 mm)

Weamer's Bucktail Body Hendrickson Emerger

page 188

Description

Nymph: (Swimming) Reddish-brown tails (3), reddish-brown-black body, legs with black banding on the tibia, and dark gray to black wing pads. **Dun:** (Male) Yellow-tan tails (3) with light brown bands. Reddish-gray body with lighter ribbing on the thorax. Reddish light gray to yellow legs. Medium gray wings, sometimes with light brown venations along the front edge. (Female) Tannish gray tails (3) with a dark gray band, yellowish-tan body and legs with a pink or olive cast, and medium gray wings. **Spinner:** Tannish-gray to dark brown tails (3) with lighter bands, reddish-brown legs and body (body has tan ribbing), and clear wings with pale brown venations. Both male and female Hendrickson spinners resemble Quill Gordon spinners. Female Hendrickson duns resemble the Sulphur *E. invaria*.

Emergence

Hendricksons begin hatching when forsythia is fully open, dandelions are blooming, and redbuds are starting to bloom. They usually continue for about two weeks, hatching in the afternoon, often during cold, rainy weather. Warmer weather often produces the best hatches, which also begin in the afternoon but may continue until dark.

Behavior and Tactics

Nymphs are poor swimmers and usually drift freely in faster currents, but they are capable of limited swimming in slower water. Dead-drift brownish-black Hare's Ears or Pheasant Tails (nymphs darken before they emerge) through riffles or slowly strip them toward slack water around in-stream obstructions (like large boulders) and toward the shore.

Hatches can be heaviest near runs at the heads of pools. Colder weather inhibits the hatching duns, and many struggle to escape their nymphal shucks, drifting downstream a long distance before they are capable of flying. Bucktail Body and Hendo Emergers, as well as dun patterns with trailing shucks, work well at this time.

Though Hendrickson hatches usually begin with a few males, it's common, later in the hatch, for both male and female Hen-

Above: Mature Hendrickson nymph; *Above right:* Female Hendrickson dun; *Right:* Male Hendrickson spinner

dricksons to hatch simultaneously. Dry flies that have bodies with a pinkish cast usually work fine for both. Sometimes more males hatch than females (or vice versa), and trout may selectively feed on one or the other. Fly patterns tied with grayish-red bodies to imitate the male (#16) and yellowish- or pinkish-tan for the female (#14) are occasionally necessary when this occurs. During particularly warm weather, often when they have been hatching for more than a week, Hendrickson hatches diminish to a trickle in the late afternoon and then a second, heavy wave of duns emerges before dark. Hendricksons provide some of the first reliable evening spinnerfalls of the season. Spinnerfalls are heavily influenced by air temperature, and while they usually occur in the early evening, late-evening spinnerfalls are common as the hatch progresses.

Notes

Hendricksons live in a wide variety of aquatic habitats. They are one of the most common mayfly species in New York. They received their common angling name from legendary Catskill Mountain fly fisherman Roy Steenrod, who designed a winning dry-fly pattern to imitate this hatch. After several seasons testing the fly on the Beaverkill River, Steenrod named the fly after his friend and angling companion A. E. Hendrickson.

Hendrickson dun body colors can vary greatly in individual streams, perhaps more so than any other mayfly species except Sulphurs. Sample duns from your local waters to find the most effective shade for your imitation. The *Ephemerella subvaria* species is believed to have several subspecies (legendary fly angler Al Caucci identifies several of them as "Subvaria X"), which have varied coloration and emerge outside the standard time frame for Hendricksons.

You can tell the difference between male and female Hendrickson nymphs by looking at their eyes. Male dun eyes are large and bright red like cherry tomatoes. Female dun eyes are smaller and olive to grayish black.

Black Quill

Male Black Quill dun

Weamer's Black Quill BB Parachute

page 189

Leptophlebia cupida, prev. *Blasturus cupidus*
Western NY: fourth week of April
Catskills: first week of May
Adirondacks: third week of May
Duration: one month
afternoon
#12–14, 2XL (9–12 mm)

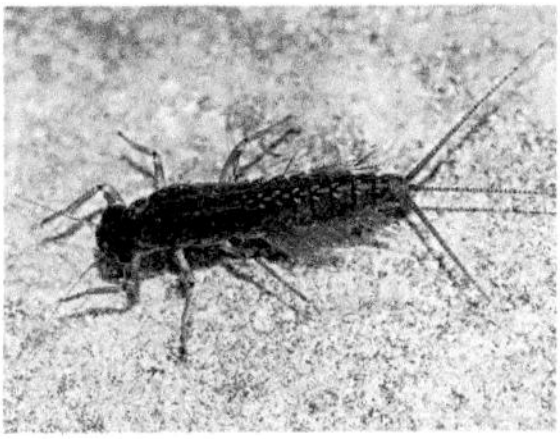

Black Quill nymph

Female Black Quill spinner

Description

Nymph: (Swimming) Tan to light brown tails (3) with dark brown banding; dark brown body with cloverlike tan marking on the back. Prominent feathery gills along each side of the abdomen. Legs are tan to light brown and wing pads are dark brown to black. **Dun:** Dark gray tails (3) with lighter bands (the dun's middle tail is shorter than its outer tails and much thinner in diameter); dark slate-gray to black body with brownish reflections and light brown ribbing; medium slate-gray wings; brownish-black front legs and tan rear legs. **Spinner:** Dark grayish-brown tails (3); dark reddish-brown front legs and light brown rear legs; coachman brown to black body with pale brown ribbing; and clear wings with very light gray and tan venations. Resembles *Isonychia bicolor, Ameletus ludens,* and *Epeorus pleuralis.*

Emergence

These large insects often begin hatching a day or two after the Hendricksons have ended, but before the start of the *invaria* Sulphurs, though it's common to find both Hendricksons and Black Quills emerging at the same time. Hatches usually occur in the afternoon when the apple trees are heavily budded and redbud and dogwood are in bloom.

Behavior and Tactics

Nymphs don't often provide consistent action before or during the hatch, because they are usually found in very shallow water before they emerge and are generally most active right before dark. But it can be effective to slowly strip dark brown Hare's Ear Nymphs along the shore and in the tails of pools. Nymphs gather in shallow stream sections and collectively migrate upstream, along the shore, often crawling long distances before emerging. Nymphs prefer to emerge into duns in flat, glassy water, where they'll undulate to and from the surface as they try to break out of their nymphal shucks. During warmer weather, fish wet flies tied to imitate the nymphs in the tails of pools and in slack areas near the shore. Let them dangle in the current at the end of the swing to imitate nymphs undulating beneath the surface.

Afternoons when the air temperature is abnormally cool or there is a fine drizzle or mist in the air usually provide the best opportunity to find trout rising to Black Quills as the inclement weather slows the duns' flight from surface. Dead-drift low-riding dry flies such as standard and Bucktail Body parachutes and Comparaduns to rising trout, which you can usually find in the tails of pools. Blind-casting dry flies is seldom effective for this hatch. Spinners (typically falling around noon) are usually more numerous than emerging duns and may provide the best fishing opportunities.

Notes

Black Quill nymphs live in slow-moving water with lots of decaying plant material, habitat that isn't prevalent in all trout streams. Where they are found, trout eat them, and it's a good idea to keep a few imitations in your boxes if the streams you frequent have even minor populations. You might not encounter a fishable hatch for several seasons, and then one day the necessary conditions align, and you're standing in the middle of trout rising to a Black Quill hatch or spinnerfall.

Most of the duns emerge in a week to ten days, but others remain as nymphs. These stragglers slowly emerge over the next several weeks, adding to the Black Quill hatch duration. I've not found the few duns that emerge outside of the concentrated hatch to be important for fishing, but you can find Black Quills for nearly a month after the first duns appear.

Big Sulphur

TCO CDC Sulphur Emerger

page 189

Ephemerella invaria, prev.
Ephemerella rotunda
Western NY: late in first week of May
Catskills: second week of May
Adirondacks: late in third week of May
Duration: three weeks
evening
#14–16 (6–10 mm)

Description

Nymph: (Swimming) Brown tails (3), reddish-brown to olive-brown body and legs, and dark gray to black wing pads. **Dun:** Light gray tails (3) with darker gray banding; creamy-yellow body and legs, sometimes with an olive or deep orange cast; darker orange to brown splotches on the back; and pale to medium gray wings. **Spinner:** Light gray tails (3) with darker banding, orange-tan body with

Big Sulphur nymphs

Male Big Sulphur spinner

yellow legs, and clear wings. Resembles many *Heptagenia* species and *Ephemerella septentrionalis*. Female duns look like *E. subvaria*.

Emergence

The hatch commences when lilacs are in bloom and a few days before March Browns begin hatching. Duns often appear in the afternoon for the first few days, but as the hatch progresses (it lasts several weeks), they commonly emerge around 8:30 PM, sometimes coinciding with a spinnerfall. At the end of their emergence for the year, hatches overlap with *E. dorothea dorothea*.

Behavior and Tactics

Most of the year, nymphs are light shades of amber or brown, but they darken significantly before their emergence. Dead-drift amber to light brown Hare's Ears through riffles to imitate the nymphs before they emerge. Fish Pheasant Tails near areas adjacent to slack water—near the shore and around large, in-stream structures like boulders—to imitate nymphs migrating from riffles to slower water to emerge into duns in or just beneath the surface film. Often nymphs struggle to escape their shucks, and they drift helplessly in the current. Patterns with trailing shucks, such as a CDC Emerger, which imitates a cripple, work well. During cold or rainy weather, flush-riding dry flies work well, and in warmer, drier weather, try hackled patterns such as Catskill-style dry flies.

Male spinners gather over riffles at dusk and wait for females to arrive. After mating, males often leave the stream before dying, depriving trout of the opportunity to eat them. Females also leave the stream, flying to nearby vegetation to wait for their eggs to ripen. But once their eggs are ready, the females return to the water, deposit their eggs, and fall spent to the water's surface. *Invaria* spinners fall in the evening as duns are emerging, making it difficult for anglers to know which stage the trout are eating.

Notes

The Big Sulphur is one of the most common mayflies in New York because it lives in a wide variety of aquatic habitats, including those degraded by pollution. This species is one of the first mayflies to return to polluted streams that are rehabilitating, suggesting that *E. invaria* has a higher water-pollution tolerance than most mayflies. Color and size can vary greatly from stream to stream.

Male dun eyes are large and bright red like cherry tomatoes. Female dun eyes are smaller and often mottled olive colored. Big Sulphur nymphs look quite similar to Hendrickson nymphs.

March Brown

Haystack (Fran Betters)

page 189

aka Gray Fox, Great Red Spinner
Maccaffertium vicarium, prev.
Stenonema vicarium, S. fuscum
Western NY: second week of May
Catskills: third week of May
Adirondacks: late in third week of May
Duration: six weeks
afternoon/early evening
#10–14, 2XL (12–16 mm)

Description

Nymph: (Clinger) Tan to reddish-brown tails (3), reddish-brown to dark blackish-brown body with an amber abdomen and darker ribbing, dark brown to black wing pads, and dark reddish-brown legs with two dark bands on the femurs. **Dun:** Amber to dark brown tails (2) with darker banding, creamy-tan belly with darker brown ribbing on back, pale to tannish-yellow wings heavily mottled with dark brown to black venations, cream legs with two dark brown bands on the femurs. **Spinner:** Dark brown tails (2) with lighter banding, tan body ribbed with dark brown, clear wings with prominent black venations and a brown area along the front tip,

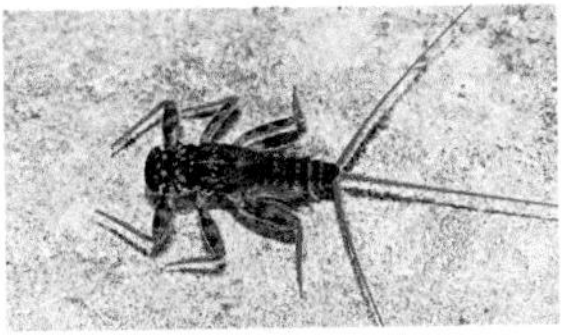

Above: March Brown nymph; Above right: Male March Brown spinner; Right: Male Gray Fox dun

amber legs with two dark bands on the femurs. Male spinners have especially large eyes, often dark brown, but sometimes colored powdery shades of blue, cream, or light gray after they've been exposed to light. Resembles *Maccaffertium ithaca.*

Emergence

The hatch begins when dame's rocket (pink and white flowers that look like phlox, but but with four petals instead of five) is just starting to bloom, lilacs have been in bloom for about week, and the Big Sulphurs are emerging. March Browns emerge during the afternoon and evening for about two weeks and then slow to a trickle for another week or two. Especially warm weather can trigger morning hatches.

Near the end of the first March Brown emergence, smaller, paler versions continue to hatch in the evenings until as late as mid-July. This mayfly used to be called a Gray Fox, but entomologists have determined that they are actually just a continuation of *Maccaffertium vicarium.*

Behavior and Tactics

Nymphs live in riffles and glides for most of the year. Dead-drift amber or brown Hare's Ears through riffles before the hatch begins. Many tiers create the wide, flat shape of the natural in their nymph patterns by squeezing a lead-wire underbody with pliers to flatten it and then dubbing over the lead. The nymphs often migrate during high-water flows from their fast-water homes to areas of slowed current to emerge into duns in or just below the water's surface. This migration may begin as early as a week before duns first appear on the surface. But if the water is low, it's common for them to emerge in riffles. During the hatch, slowly strip Beadhead Hare's Ears along the stream bottom toward areas of slack current to imitate naturals preparing to emerge into duns.

Imitate emergers and duns with deer hair emerger patterns, Haystacks, or flush-floating dry flies dead-drifted to rising trout or blind-casted near pockets of slack water in riffles and runs. Use Catskill-style flies or other patterns with hackle to imitate duns that are aggressively flapping their wings before flight, common with March Browns.

Spinners often appear in massive numbers, the combined total from several days' worth of duns. It's common for spinners to hang over riffles at dusk, giving the appearance that they may fall at any second. Then suddenly they are gone, leaving puzzled anglers. I spent many nights collecting March Brown spinners and cataloging the swarm's male-to-female ratio (remember, males hatch first). On the nights when the spinnerfall disappeared, I captured very few female spinners, usually none at all. I believe that the March Brown spinnerfall usually manifests itself just like the hatch, in a trickle. Males hatch, trickling for days even a week, before most females emerge. Each night more males join the spinner swarm, but there aren't enough females. So the few females mate, forcing the bulk of the male swarm to return to the trees to wait for the next night. One night, the male-to-female ratio will be at its closest level, and the best spinnerfall of the year will take place. Years where climatic events cause both males and females to hatch simultaneously create conditions for epic March Brown spinnerfalls.

Notes

March Browns are especially prevalent in streams and rivers with plenty of fast water and rocks. The hatch often trickles throughout the day, seldom producing large numbers of flies, but the fish remain focused on the surface, keenly aware of the possibility of a March Brown dun entering their feeding lane. Because of this, the hatch provides the first opportunity of the season to blind-cast dry flies through riffles, runs, and other places that look like they might hold fish.

The Haystack tied by Fran Betters may look a little messy, but it is a very effective fly and the predecessor of Caucci and Nastasi's Comparadun. Mr. Betters was a friend (sadly, he passed away in 2009) and one of the Adirondacks' most important fly designers.

Notice the smaller abdomen and the faded coloration of the male Gray Fox dun compared to the male March Brown dun.

Little Sulphur

Biot Body CDC Comparadun

page 190

aka Pale Evening Dun, Pale Watery Dun, Summer Sulphur

Ephemerella dorothea dorothea, prev. *Ephemerella dorothea*

Western NY: third week of May

Catskills: fourth week of May

Adirondacks: first week of June

Duration: three weeks (longer in tailwaters)

evening

#18–20 (6–8 mm)

Little Sulphur nymph

Female Little Sulphur spinner

Description

Nymph: (Swimming) Pale olive to ginger tails (3) with dark brown bands, dark brown body with amber areas, dark brown mottled wing pads, pale olive to ginger legs with dark brown markings. **Dun:** Cream to pale light gray tails (3) without banding (*invaria* or Big Sulphurs have bands on their tails); pale creamish-yellow to orange body, sometimes with olive reflections and no, or few, brown patches on the back (*invaria* Sulphurs have brown markings on their backs); light gray wings; pale yellowish-cream legs. **Spinner:** Cream to light pale gray tails (3), creamish-yellow belly, darker orange on back, glassy clear wings, and pale cream legs. Resembles *Ephemerella invaria, Ephemerella septentrionalis, Leucrocuta aphrodite,* and *Leucrocuta hebe.*

Emergence

Little Sulphurs usually begin hatching during evenings when rhododendron flowers are opening and mountain laurel is in bloom. They continue for several weeks, overlapping with other mayfly species, including Green Drakes.

Behavior and Tactics

Before the hatch begins, dead-drift Beadhead Pheasant Tails through riffles or suspend them beneath a dry fly in slow-moving pools. Poor swimmers that live in nearly all types of aquatic habitat (from slow pools to riffles), Little Sulpher nymphs can emerge into duns in the surface film, but they mainly float just beneath the surface, wiggling out of their nymphal shucks. To imitate this, fish unweighted Pheasant Tails, greased and fished in the film, with a dead drift and an occasional twitch to bulging fish (fish that are eating with boiling rises just beneath the surface).

The heaviest hatches most commonly occur at dusk or dark and bring many trout to the surface. When trout are actively feeding on Little Sulphurs in calm, flat pools, dead-drift emerger patterns and flush-floating dry flies such as Biot Body CDC Comparaduns. Once *E. dorothea dorothea* has been hatching for a few days, spinners fall at the same time the duns are hatching. When this happens, try a dun imitation first, since they are easier to see in low light. At times, however, the trout feed selectively on spent-wing patterns.

Notes

In some tailwater fisheries, like the Upper Delaware River, *E. dorothea dorothea* produces more than one generation per year, and they often emerge in the afternoon even during the summer heat. This must be caused by the artificially cold temperatures created by the dam's water releases. Hatches on the Upper Delaware can last as long as three months, but the size of the duns decreases as the hatch progresses.

Little Sulphurs commonly have wide variations in body color, depending on the stream and what the nymphs ate. Little Sulphurs do not reside in as many streams as Big Sulphurs (*Ephemerella invaria*), but they are still fairly common throughout the state.

Notice the lack of prominent banding on the Little Sulphur dun's tail compared to the Big Sulphur dun. Little Sulphurs also often lack any darker markings on their backs. Many anglers have trouble telling if they're looking at Big or Little Sulphurs when they capture one. But the much smaller, paler, bland-looking Little Sulphur looks significantly different when the two flies are seen side by side.

Little Sulphurs have two redundant names for their species: *E. dorothea dorothea*. This is called a trinomial, and it's used because they have been joined as a subspecies with the Western species *E. dorothea infrequens*.

Green Drake

Female Green Drake dun

Weamer's Green Drake Comparachute

page 191

aka Shad Fly, Coffin Fly
Ephemera guttulata
Western NY: third week of May
Catskills: fourth week of May
Adirondacks: late in first week of June
Duration: one week
evening
#8–12, 2XL (17–29mm)

Dette Coffin Fly (Mary Dette)

page 190

Description

Nymph: (Burrower) Dark olive-gray tails (3); two thirds of the body is creamy tan but the final few tergites (body segments), closest to the tail, are dark olive-gray; dark brown to black wing pads; and creamy tan to ginger legs. Prominent, olive-gray feathery gills that curl over the top of the back when the fly is at rest. **Dun:** Dark brownish-black tails (3), pale yellowish-cream belly with olive-gray back, yellowish-green wings with black venations and blotching, dark olive-black front legs and cream rear legs. Male duns are often much smaller (one to two sizes) and darker in color than females. Males often have dark olive-gray to black sleeves on their front legs. **Spinner:** Light brown tails (3) with dark brown bands; body is creamy off-white to pale yellow (with eggs) or chalky white (after depositing eggs); clear wings (sometimes with a yellow cast) with prominent black venations and blotching; dark brown to black front legs and off-white back legs. Resembles *Litobrancha recurvata, Ephemera varia, Hexagenia rigida.*

Emergence

Eastern Green Drake emergences begin when the white flowers on black locusts are almost fully open. Odd, unseasonably warm or cool weather has made timing the hatch more difficult in recent years, making contact with local anglers or a fly shop near the water vital for timing the Green Drake emergence on any fishery. Sensitive to light, Green Drakes often begin hatching in the last hour before dark. On cloudy days or on streams shaded during the day by mountains or heavy tree canopy, hatches can begin in the afternoon and continue into dark.

Behavior and Tactics

Nymphs burrow into fine gravel and sandy areas in the streambed. When they leave their underground homes to feed or emerge, they become vulnerable to trout. Before the hatch begins, dead-drift or slowly strip large, heavily weighted tan Hare's Ears through riffles. Green Drake nymphs swim to the surface to molt into duns, wiggling up and down at or just beneath the surface as they try to shed their nymphal skins. This struggling movement is irresistible

Mature Green Drake nymph

Male Green Drake spinner (Coffin Fly)

to trout but difficult to imitate with fly patterns. Try fishing an unweighted or lightly weighted Hare's Ear or wiggle nymph with short, quick strips or swinging the fly in the current, allowing it to dangle at the end of your presentation so the flowing water moves the fly and mimics the natural's up-and-down motions.

Many anglers leave the stream too early and then complain that this hatch is overhyped. It's common, especially during warm weather, for duns to emerge in large numbers only after dark. Early on, the duns are best imitated with large dry flies, such as Comparachutes, with dark brown trailing shucks. But after the duns have been emerging for more than a couple of days, trout tend to prefer duns that move (at least during the day). It's difficult to mimic the way that the large Green Drakes flutter their wings before flight, and trout can easily discern a forgery. A high-riding Catskill-style pattern gently twitched about a foot in front of a rising trout and then fished drag-free over the fish sometimes works well.

The peak of the hatch, and sometimes the best fishing, occurs during the spinnerfall. About an hour before dark, the spinners usually flutter in and out of the treetops—sometimes in such staggering numbers on some rivers that you can hear their wings clicking as they fly. As darkness approaches, massive waves of Coffin Flies fly upstream looking for mates. After mating, the males die and fall spent to the water. The females deposit their eggs and also fall spent.

The intense spinnerfalls cause two major problems for anglers. First, this extreme biomass of food on the water often makes it difficult to get trout to notice your fly. Second, because much of the action takes place after dark, it is hard to see your fly to determine whether you are casting accurately or if a fish has taken it. Using an easy-to-see pattern such as a #4-6 Hi-Vis White Wulff or a Dette Coffin Fly (tied on 2X-long hooks) helps solve both of these problems. These oversized patterns are easier for the fish to see, and you can also track them in the moonlight. Cut back your tippet to 3X or 4X to prevent the larger flies from twisting your tippets and to give you the extra strength for landing the larger trout that feed after dark.

Notes

The nymphs have a two-year life cycle, so duns that emerge this year have hatched from eggs that were deposited two seasons ago. The common angling name for the Green Drake spinner, the Coffin Fly, was posed by Catskill Mountain fly-fishing legends Walt Dette and Ted Townsend. They coined the name after a funeral, while tying flies to imitate the chalky white spinners.

Sulphurs usually emerge at the same time as Green Drakes, and trout often prefer this smaller fly. Study a rising fish closely to determine whether it is eating Sulphurs or Drakes. But even if the fish you are casting to is obviously eating Green Drakes, it's a good idea to show it a Sulphur pattern if it has refused your Drake imitations. Slate Drakes also often appear around the same time as Green Drakes, an hour or two earlier in the evening.

Size varies greatly from stream to stream, but males are usually one to two sizes smaller than females.

All of the Green, Brown, Dark Green, and Yellow Drake nymphs, as well as Hex species nymphs, look similar because they are all burrowing mayflies. From a fly-fishing perspective, you don't need to be able to tell the difference between these nymph species. The same nymph patterns and fishing techniques can be applied to all of them.

Dark Green Drake

Female Dark Green Drake dun

Weamer's TF Comparadun,

page 190

Litobrancha recurvata, prev. *Hexagenia recurvata*
Western NY: third week of May
Catskills: fourth week of May
Adirondacks: second week of June
Duration: approximately one week
afternoon
#6–10, 2XL (20–38 mm)

Description

Nymph: (Burrower) Brown outer tails (2), cream to ginger middle tail, reddish-brown body with tan to cream sides and belly, dark brown wing pads, prominent olive-brown feathery gills, and ginger legs with darker brown areas. **Dun:** Long, brownish-black tails (2) with a miniscule vestigial third tail between them; dark slate-olive-gray body prominently ribbed with yellow; dark gray to creamy-tan wings with olive-green and reddish-brown reflections and prominent light brown venations; dark reddish-brown front legs and olive-brown rear legs. **Spinner:** Brown tails (2); orangish-tan to bright, deep red-orange body; pale gray to clear wings with areas of black and brown shading and prominent light brown venations; dark reddish-brown legs. Resembles *Ephemera guttulata.*

Emergence

Duns hatch when the rhododendrons are just opening, often in the afternoon hours, and they continue hatching for about one week.

Behavior and Tactics

Nymphs and duns behave just like Green Drakes, so fish Dark Green Drake imitations in a similar manner. Spinners usually fall at dusk and can be heavy if cold weather forces several days' worth of spinners to fall in the same evening. It's easier to catch fish during Dark Green Drake spinnerfalls than it is during Green Drake spinnerfalls because there are usually fewer naturals on the water.

Notes

On some small brook trout streams, the Dark Green Drake may be one of the only large meals of the season. Because duns are so large, and often emerge in early afternoon, they usually live in small streams with sandy or silt-bottomed areas and tight, heavy tree canopies. The vegetation allows them to hide from predators quickly after hatching. Like *Ephemera guttulata,* nymphs have a two-year life cycle, and the duns that emerge this year have hatched from eggs that were deposited two seasons ago. The Dark Green Drake dun may be the most visually stunning of all mayflies. The

Mature Dark Green Drake nymph

Female Dark Green Drake spinner

dun's dark body is dramatically ribbed with yellow and its wings look like stained glass windows. And the spinners are equally spectacular, with deep, bright orange bodies, rivaling the beauty of any mayfly in North America.

Often Dark Green Drakes and Green Drakes appear concurrently. Though they appear similar to Green Drakes, the two species are unrelated. Dark Green Drakes have two long tails with a miniscule, vestigial third tail between them (the short vestigial tail is a common characteristic of *Hexagenia,* a genus that once included the Dark Green Drake). Green Drakes have three equally long tails. Also, Dark Green Drakes have dark-colored bellies, but Green Drake bellies are light colored.

The easiest way to differentiate Green Drake nymphs and Dark Green Drake nymphs is to look at their frontal processes. The frontal process is the liplike structure that protrudes from the front of a burrowing mayfly's head. Green Drakes have a deeply cupped frontal process while the Dark Green Drake's is slightly rounded.

The female Dark Green Drake is the largest mayfly in North America, rivaled only by *Hexagenia limbata* females.

The wings of the female spinner in the accompanying photo are frayed along the upper edges. I've seen many photos of Dark Green Drake spinners with similarly frayed wings. The flies are so large that they rub the tops of their wings against obstructions (like the lid of the Tupperware container I used to watch this dun to transform into a spinner), though the fraying doesn't inhibit flight. This one flew away just after I pressed the shutter to capture this photograph.

Slate Drake

Female Slate Drake dun

Weamer's Burke Emerger,

page 190

aka Leadwing Coachman, White-Gloved Howdy, Iso
Isonychia bicolor, prev. *Isonychia harperi, matilda, albomanicata, sadleri*
Western NY: third week of May, second week of September
Catskills: fourth week of May, second week of September
Adirondacks: second week of June, third week of August
Duration: one month (first generation), two weeks (second generation)
evening, afternoon
#10–14, 2XL (12–17 mm)

Description

Nymph: (Swimming) Dark gray to orangish-brown tails (3), lighter at their tips; reddish-brown to dark brown to dark maroon body with a lighter lateral stripe on the back; prominent, tear-shaped gills along the sides of the abdomen; dark gray to black wing pads; pale

Slate Drake nymph

Male Slate Drake spinner

olive-gray to tan legs with dark banding on the femur and tibia. **Dun:** Medium brownish-gray to light gray tails (2), reddish-mahogany or dark slate-gray body with olive reflections (depending on the stream). Some have bright green bodies when they first emerge that quickly fade to olive-gray as they are exposed to air; front legs match body color, rear legs are cream. Wings are medium gray, occasionally with a lighter patch in the center. **Spinner:** Medium to light gray tails (2), deep red to reddish-mahogany body, clear wings, rear two pairs of legs are cream and front pair is dark brown. Resembles Dark Quill Gordon (*Ameletus ludens*) and Gray Drake (*Siphlonurus quebecensis*).

Emergence

Isonychia bicolor produces two independent hatches each year. The first hatch appears when mountain laurel is blooming and oxeye daisy is beginning to bloom. The second appears when aster is blooming. These are not believed to be two broods, the second being descendants of the first, but rather two separate hatches much like the every-other-year Green Drake emergence. It's also common to find a few Isonychia hatching throughout the summer months, particularly during cool or rainy weather. Duns often emerge around 7 PM during warm weather in the late spring. Afternoon emergences are common during the cooler weather that often accompanies the second hatch. It's common to see March Brown spinners when the first hatch appears and Little Blue-Winged Olives with the second.

Behavior and Tactics

To imitate the great swimming ability of the nymphs, strip Prince Nymphs or Pheasant Tails through riffles toward the shore with short, quick tugs on the fly line, or cast weighted wet flies tied with peacock-herl bodies and reddish-brown hackle collars into fast water and retrieve them toward shore with the same stripping techniques. Picket Pin wet flies often work well for the *Isonychia* hatch. Use 4X tippets, as fish can strike violently.

Nymphs live in fast water and swim or crawl to rocks and debris along the shore to emerge, though some do emerge in the surface film.

It's uncommon to find a lot of *Isonychia* duns on the surface, but trout seem to be on the lookout for those that hatch in-stream, making Isonychia dry flies and Burke Emergers excellent searching patterns when *Isonychia* are hatching.

Springtime spinnerfalls usually occur at dusk, but also in the afternoon during cool weather. Afternoon falls are common during the second hatch, in the autumn. The spinners can provide excellent dry-fly opportunities, but *Isonychia* hatch in the same manner as March Browns—in a trickle—and it's common for their spinners to fall in a trickle as well, just as I described in the March Brown section. Imitate evening *Isonychia* spinners with Poly Wing Spinner patterns to aid your ability to see them. For afternoon spinners, sparse hackle-wing spinners are often a better choice.

Notes

A good indication that the hatch has begun is the presence of black nymphal shucks on rocks along the edges of the stream. Size varies from hatch to hatch and stream to stream. Duns produced during the first hatch, May through June, are larger (#10–12) than the duns in the second hatch, September through October (#12–14). Body color also varies. Freshly emerged duns sometimes have a bright green or olive coloration to their bodies—it's important to have a few emerger patterns tied with these body colors. The flies darken when they are exposed to air, so trout rising to the duns may take different-colored flies from those rising to emergers.

Generally, the second hatch produces better fishing opportunities than the first. There are far fewer other insects on the water in September and October to compete for the trout's attention.

The *Isonychia* nymph in the photograph is immature. Its wing pads lengthen and become darker, and much more prominent, as the nymph develops.

Brown Drake

Female Brown Drake dun

Weamer's Brown Drake Flip Flop Fly

page 191

Ephemera simulans
Western NY: fourth week of May
Catskills: first week of June
Adirondacks: second week of June
Duration: less than one week
evening
#10–12, 2XL (11–16 mm)

Brown Drake nymph

Male Brown Drake spinner

Description

Nymph: (Burrower) Tan tails (3); creamy tan belly (with pale brown horizontal markings) and sides; light brown feathery gills; brownish gray back; dark brown mottled wing pads; and tan to ginger legs. **Dun:** Light brown tails (3) banded with dark brown, tannish olive-yellow, or olive-brown; belly and legs with dark brown markings; brown back; tan wings with olive reflections and prominent dark brown to black venations and blotching. **Spinner:** Tails (3) tan with dark brown bands, olive-brown body with dark brown markings, clear wings heavily barred with dark brown venations and blotching, dark brown front legs and cream rear legs with dark brown tips. Nymph so closely resembles Yellow Drake *Ephemera varia* that the two can be difficult to tell apart, though Brown Drake nymphs are more developed earlier in the year. Dun resembles March Brown (*Maccaffertium vicarium*), but March Browns are usually a bit smaller and lighter in color, and have two tails. Spinners resemble *Hexagenia atrocaudata* spinners, but the latter is larger and has two tails rather than three. (*Hexagenia* has a vestigial middle tail.)

Emergence

Brown Drakes hatch when peonies are in full bloom and the flowers on the black locust trees are almost fully open. Like Green Drakes, Brown Drakes are sensitive to light and emerge at dusk or later, but on cloudy days or on streams shaded during the day by mountains or heavy tree canopy, hatches can begin in the afternoon.

Behavior and Tactics

Brown Drake nymphs prefer to live in slow to moderate currents where they burrow into fine to fairly coarse streambed substrate and silt. Large, heavily weighted, tannish-gray Hare's Ears catch fish if they are dead-drifted or slowly stripped along the streambed in the braided water below riffles before the hatch begins. Brown Drake nymphs swim to the surface with an up-and-down undulating motion like Green Drakes. They emerge at, or just below, the surface. Fish the imitations in a similar manner to Green Drakes (see page 48).

Male spinners begin their mating flight near the trees around 7 PM. Female spinners, after mating, fly toward the surface and travel upstream. More than any other mayfly species, Brown Drake duns and spinners are concentrated for only a few days on a stream, making spectacular spinnerfalls common. Often, your fly patterns have to compete with thousands of naturals for the trout's attention. Try using oversized flies (one to two sizes larger than the natural) to get the trout's attention and help you spot your pattern, sink your dun or spinner a few inches beneath the surface, or use spinner patterns with upright wings—Brown Drake spinners have the unusual tendency of sometimes floating with their wings held upright after mating instead of being spent or flat on the water. You can also fish a spinner pattern behind a highly visible dun imitation.

Notes

Because Brown Drakes begin as the Green Drake hatch is slowing, the number of Brown Drake duns compared to Green Drake duns is a good indicator of how much longer the Green Drakes continue to hatch in streams that contain both mayflies.

Brown Drake nymphs look very similar to Green Drake nymphs. The easiest way to tell them apart is to look at their bellies. Brown Drake nymphs have light brown horizontal lines on their bellies and Green Drakes don't. The frontal process (see Dark Green Drake, page 52) is also shallower on Brown Drake nymphs, though this is difficult to see streamside, without some sort of magnification.

Blue-Winged Olive

Male Blue-Winged Olive dun

Hackle Wing BWO

page 191

aka BWO, Cornuta

Drunella cornuta, D. cornutella, D. lata. prev. *Drunella lata* for all three species, *D. longicornis, Ephemerella cornuta, E. cornutella, E. lata*

Western NY: first week of June (slightly later for *D. lata*)

Catskills: second week of June (slightly later for *D. lata*)

Adirondacks: third week of June (slightly later for *D. lata*)

Duration: ten days (several weeks for *D. lata*)

late morning/early afternoon

#14–16 (6–10 mm, #16-18 for *D. lata*)

Description

Nymph: (Swimming) Brown tails (3), lighter at the tips; brown body peppered with faint darker brown dots along the thorax and tan ribs across the abdomen; mottled brown to black wing pads; and medium brown legs. **Dun:** Pale gray to gray-olive tails (3);

Immature Blue-Winged Olive nymph

Female Blue-Winged Olive spinner

bright insect-green to medium olive body (depending on how much time has lapsed since the duns hatched); medium to dark slate-gray wings; amber-olive legs. **Spinner:** Gray tails (3), blackish-olive body, clear wings, dark olive-brown legs. Resembles other Blue-Winged Olive species like *Drunella attenuata, Drunella walkeri*, and *Ephemerella needhami* (female).

Emergence

Hatches begin when peonies are opening. Emergences often occur late in the morning and early in the afternoon, though weather can influence this timing. Evening emergences and spinnerfalls are also relatively common.

Behavior and Tactics

Dead-drift weighted Pheasant Tails through riffles, or slowly strip them from fast water to slack water to match *Drunella* nymphs preparing to emerge. Even though they live in riffles and sections of streams with current, *Drunella* nymphs are poor swimmers, so they migrate to calmer water to emerge in or below the surface. Unweighted Pheasant Tails greased and cast to rising fish, or slowly twitched and stripped below the surface, imitate nymphs emerging into duns.

Brightly colored when they first emerge, duns darken when exposed to air. They often escape from the surface quickly, so evening spinnerfalls sometimes produce better fly-fishing opportunities than the hatch. If you hit this hatch on a cool, overcast, or drizzly day, the duns stay on the water longer, and trout feed heavily on emergers and duns. Patterns tied with bright green bodies imitate the color of newly hatched *Drunella*. Trout feed heavily on *Drunella* spinnerfalls, so be prepared with black-bodied patterns that have exaggerated thoraxes to imitate the spinners' unusually robust olive-black bodies.

Notes

Typical hatches begin in late morning, and they continue for an hour or two. But extremely warm weather can cause a hatch to occur earlier in the morning. Especially cool weather has the opposite effect, delaying an emergence until after 12 PM. Weather can also affect the spinnerfall, which occurs in the evening after 7 PM. If the weather is warm, spinners often won't reach the water until after dark or early the next morning.

For less than a year, the *Drunella cornuta* and *Drunella cornutella* species were reclassified and merged with *Drunella lata*. The timing of this reclassification created an error in the *Pocketguide to Pennsylvania Hatches*, a book that I wrote with Charlie Meck. When the two species were separated and returned to their original classification, our book was already printed and it was too late to make a correction. Entomology is always changing as advancements in aquatic insect study lead to new understanding of species relationships. And it's strangely comforting to know that sometimes even professional entomologists make errors.

D. cornuta, D. cornutella, and *D. lata* look and behave similarly enough, and can often be found in the same stretches of rivers and streams, so exact species identification is not important for the fly fisher. Fly patterns and tactics for one species work well enough for all of them. A paper by David Funk, Bernard Sweeney, and John Jackson listed in the bibliography of this book has some fascinating details about the three species for those who are interested in gaining a greater understanding of the similarities and differences between them.

The glossy sheen on the *Drunella* spinner lasts only a short time after the fly has molted from a dun. I shot the photograph here seconds after the fly changed to a spinner. This female spinner's body darkens significantly and its abdomen shrinks before it falls spent to the water after laying its eggs.

Gray Drake

Female Gray Drake dun

Usual (Fran Betters)

page 191

Siphlonurus sp.
Western NY: first week of June
Catskills: second week of June
Adirondacks: second week of June
Duration: heaviest for two weeks, sporadic for one to two months
#12–14 (13–16 mm)

Description

Nymph: (Swimming) Olive-tan tails (3) with dark bands near the tips; olive-tan to olive-brown body with reddish-brown markings and prominent black veins across the back; tear-shaped gills; mottled brown to black wing pads; and tannish-olive legs with prominent black bands. **Dun:** Pale gray tails (2) banded light brown, slate-gray body with lighter areas, watery gray wings with tan venations, olive legs with areas of amber. **Spinner:** Light gray tails (2) with dark gray banding, dark slate body with pale tan blotching, clear wings with pale gray venations, tan legs with amber areas. Resembles *Isonychia bicolor*.

Gray Drake nymph

Female Gray Drake spinner

Emergence

Gray Drakes most often emerge by crawling onto streamside vegetation, often in the afternoon hours, and they seem particularly fond of grasses that grow along some stream banks, though some emerge in the surface film. But even when the flies do emerge in the water, it is usually very close to the shore. You seldom find their nymph shucks on rocks like you often do with Slate Drakes *(Isonychia)*. This emergence characteristic often makes the duns unimportant to fly fishing because trout do not have the opportunity to eat them. Like most every mayfly rule, though, there are exceptions, and I have encountered days where enough Gray Drake duns were found in the water for trout to key on them.

Behavior and Tactics

Gray Drake nymphs live in silty areas out of the main current—slack areas along shore, backwaters, and flood pools along islands. They are excellent swimmers and move through the water like minnows. Because the nymphs are most often found in shallow, nonmoving stream areas, trout have little opportunity to eat them. However, there are times, often after dark, when trout move into Gray Drake nymph habitat to feed. Fishing olive-gray Hare's Ears in slack areas adjacent to current can produce a few fish. Work the flies with quick, short strips to imitate the swimming nymphs. Duns can be imitated by an Adams or a gray-bodied Usual. Look for trout rising to Gray Drake duns very close to shore. Spinnerfalls often occur at dusk, occasionally in the morning, and they usually provide the best fly-fishing opportunities. A gray-bodied Antron or hackle-wing spinner works well.

Notes

Gray Drakes look similar to Slate Drakes at a distance. The easiest way to tell the duns apart is to look at their legs. Slate Drake legs look thicker and the rear two pairs are cream while the front pair is dark. In comparison, Gray Drake duns have thin, spindly-looking legs that are all the same color. Gray Drakes' color can be quite variable. I've captured Gray Drake spinners that were red, almost maroon, similar to a Slate Drake spinners. The color variation could be caused by diet or perhaps genetic variations in various *Siphlonurus* species.

Gray Drakes have an extended emergence. Most of the hatch occurs in one week to ten days, and this is when you'll find the best fishing opportunities. But it's common to find occasional stragglers emerging more than a month after the main emergence has ended. The extended emergence is usually a low-density affair with few fly-fishing applications.

Light Cahill

aka Orange Cahill
Stenacron interpunctatum, prev. *Stenonema canadense, Stenacron canadense, S. interpunctatum, S. heterotarsale*
Western NY: first week of June
Catskills: second week of June
Adirondacks: second week of June
Duration: one month
early evening
#14–16 (9–12 mm)

Ausable Wulff (Fran Betters)

page 191

Description

Nymph: (Clinger) Medium ginger to light brown tails (3), yellow body with dark brown markings on its back, mottled brown to black wing pads, and cream to tan legs with darker areas. **Dun:** Creamish-yellow to light gray tails (2); orangish-yellow to deep orange (female) or creamy yellow (male) abdomen with orange thorax, often with olive reflections; yellow wings with brownish-black barring and dark edging on the hind wing; and yellow legs with two dark brown bands across the femur. **Spinner:** Light dun tails (2) and creamish-yellow-tan to yellow body (the female's abdomen has a distinct orange cast from the eggs); and glassy clear wings with faint light brown venations and dark edging on the hind wing. Resembles *Epeorus vitreus* (female only).

Emergence

A few Light Cahill duns may begin to appear sooner, but the hatch isn't usually significant until June, when oxeye daisies and mountain laurel are blooming. Emergences usually occur in evening, beginning around 7 PM. Light Cahills often appear with Green Drakes or Sulphurs, so be prepared to fish both hatches.

Behavior and Tactics

Nymphs are clingers and share the same wide, flat body as March Browns. Dead-drift tannish-brown Hare's Ears through riffles and runs before the hatch begins. Nymphs migrate to slower sections of the stream, often near the middles and tails of pools, where they drift or swim just below the surface and emerge into duns.

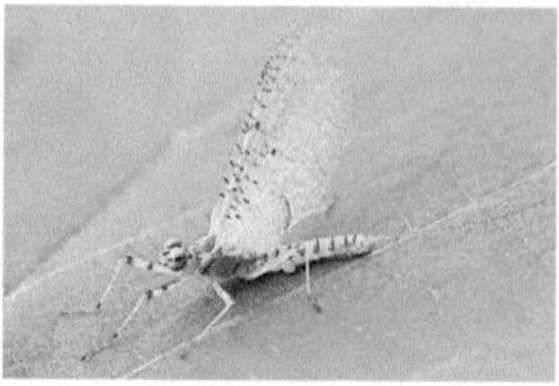

Above left: Light Cahill nymph; *Above:* Male Light Cahill dun; *Left:* Male Light Cahill spinner

Standard dry flies, tied without trailing shucks, such as Comparaduns, Ausable Wulffs, and parachutes, catch fish that are rising to duns. Trailing shucks are unnecessary because the nymphs emerge into duns below the surface, though you can incorporate them into your patterns for the same triggering effect described in the Quill Gordon section. Light Cahills hatch when the light is low, and hackled, high-riding drys are easier to see. Light Cahill spinnerfalls create some of the best dry-fly fishing opportunities for this species.

Notes

Anglers often confuse the deep orange abdomen of female Light Cahills with Sulphurs and every year someone in the fly shop tells me that they saw giant Sulphurs emerging when what they really saw were Light Cahills. Coloration varies greatly from stream to stream.

The odd growth in the middle of the male dun's abdomen in the photograph must have been caused by an injury. This is not a normal attribute of Light Cahill male duns.

Cahill

Female Cream Cahill dun

Cream Variant (Joe Fox)

page 192

aka Cream Cahill, Dark Cahill
Maccaffertium species and *Stenonema femoratum*, prev. *Stenonema* and *Heptagenia* species
Western NY: first week of June
Catskills: second week of June
Adirondacks: third week of June
Duration: three months
evening
#14–16 (8–11 mm)

Description

Nymph: (Clinger) Pale gray to amber tails (3); amber to brown body (some species heavily mottled black on back) and dark brown to black wing pads; cream to tan legs with black banding on the femurs. **Cream Cahill Dun:** Cream tails (2) barred light brown; pale cream to creamy tan body; cream to light pale yellow to light

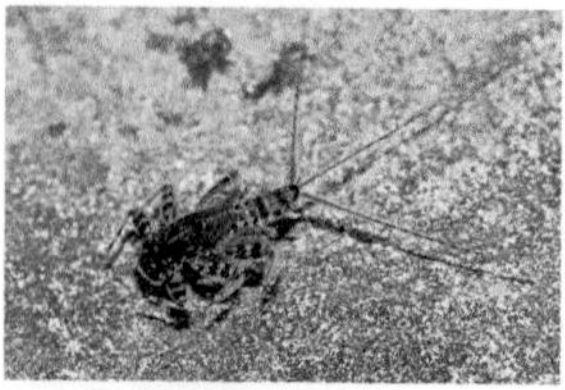

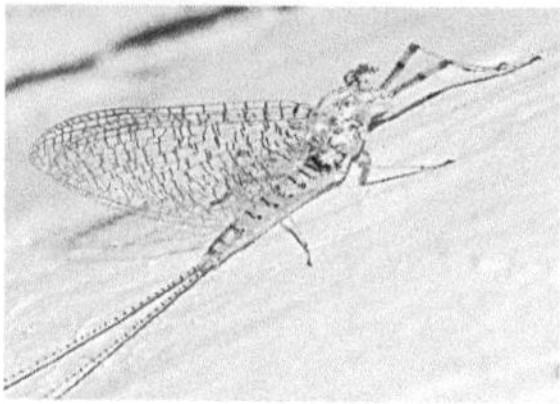

Above left: Maccaffertium nymph; *Above:* Female Dark Cahill dun (probably *M. ithaca*); *Left:* Female Cream Cahill spinner

pale gray wings with prominent but light venations. Legs are pale cream to creamish-tan with brownish banding. **Dark Cahill Dun:** Cream to light pale gray tails (2) barred light brown; tan to yellow-tan body with darker marking on the back; tan to yellow-tan legs with light brown banding. **Cream Cahill Spinner:** Cream tails (2) with light brown banding; cream or pale yellow (if it still has eggs in it) or white (if eggs are exuded) body; clear wings with light brown venations; white to pale cream legs with light brown banding. Some species resemble the White Fly (*Ephoron leukon*). **Dark Cahill Spinner:** Pale gray to light brown tails (2) with darker banding; tan body with darker brown marks on the back; clear wings with pale brown venations; yellowish-tan legs with light brown banding.

Emergence

Most Cream and Dark Cahills share many of the same physical traits and emergence characteristics as March Browns—nearly all of them are part of the *Maccaffertium* genus. Cahills begin hatching in the evening when oxeye daisies, mountain laurel, and milkweed are blooming, and a few may emerge as late as September.

Behavior and Tactics

Dead-drift amber or brown Hare's Ears through riffles before and during the hatch. Tie these flies wide and flat, like those described in the March Brown section, to imitate the distinct body shape of these flat clingers that live in swift-moving riffles and glides for most of the year. Standard March Brown nymphs, tied two sizes smaller than you would use for the *M. vicarium* hatch, also work well. Cahills emerge into duns in or just below the water's surface. They usually hatch in the early evening from 7 to 9 PM. Usually, emerger patterns are not very effective for Cahill hatches because they are hard to see so late in the day, and trout show little preference between emergers and the easier-to-see dun patterns like Variants. Hatches are not often heavy or concentrated, but the spinnerfalls at dusk can be heavy and often provide the best dry-fly-fishing opportunities. Catskill- and Wulff-style flies are good choices for spinner imitations because they are easy for you, and the fish, to see as dark approaches—important during heavy spinnerfalls. These hackled flies also imitate female spinners hovering above and briefly touching the water to release their eggs.

Notes

The angler names "Cream Cahill" and "Dark Cahill" encompass a wide range of mayfly species. Identifying each one to the species level is unnecessary for the fisherman. Because their hatches extend to late summer, anglers often confuse Cream Cahills with White Flies. But unlike White Flies, Cream Cahills fly farther above the surface when they emerge and mate. All Cream Cahills have two tails; male White Flies have three.

There are several Cahill species in the photographs, but they all complete their life cycles in similar manners. Sample your local Cahills and adapt your flies to match them.

Do not underestimate the importance of these mayflies simply because I have merged many *Maccaffertium* species together under the name "Cahills." From an angler's perspective, these are some of the most important summer mayflies.

Yellow Drake

Female Yellow Drake dun

Ephemera varia
Western NY: second week of June
Catskills: third week of June
Adirondacks: fourth week of June
Duration: six weeks (most duns hatch in approximately 10 days)
evening
#10–12, 2XL (12–16 mm)

Irresistible (the Dette shop)

page 192

Description

Nymph: (Burrower) Amber to light brown tails (3); amber body with darker brown markings on the back and fuzzy brown gills that extend from the sides of the abdomen to the top of the back when the nymph is at rest; medium brown to black mottled wing pads; and creamish-tan legs. **Dun:** Tan tails (3) with darker brown banding; pale yellow abdomen with brown markings on the belly that

Mature Yellow Drake nymph

Male Yellow Drake spinner

look like quotation marks and brown markings on its back; pale orange to peach-colored thorax and pale yellow wings with light brown venations (hind wing has few to no brown venations); pale yellow legs, front pair with darker brown banding. **Spinner:** Creamish-yellow tails (3) banded with dark brown; creamish-yellow belly with dark brown lines, especially on last four segments; back is heavily mottled with dark brown and reddish-brown markings; clear wings with dark brown blotches and venations and edged dark brown; rear two pairs of legs creamish-yellow and front legs tannish-yellow with darker markings. Yellow Drake duns look nearly identical to the spinners. The only difference is that spinners' wings are clear. Resembles *Hexagenia rigida* and *Ephemera guttulata*.

Emergence

Yellow Drakes appear when elderberry and chicory are just blooming. This hatch is especially long-lived, and occasional stragglers may emerge into August. The duns usually begin emerging in the last hour before dark—around 8:15 if the weather has been cooler, closer to 8:45 if it's been very warm—and continue hatching for approximately half an hour.

Behavior and Tactics

Yellow Drake nymphs are burrowers that prefer to live in warmer, slower sections of streams. Prior to hatching, they leave their burrows and undulate toward the surface to emerge in the film. Imitate this behavior by fishing amber and brown Hare's Ears with the same tactics as for the Green Drake nymph (page 48).

Yellow Drakes seldom hatch in large numbers, and trout sometimes ignore them because they often emerge when the water is warm. But the duns quivering on the surface as they attempt to fly can bring aggressive rises from trout. Try skittering hackled flies in front of a rising trout to imitate this. Just make sure that you're not planning to catch and release wild trout in water over 70 degrees; the fish will probably die.

Duns and spinners are often on the water right before and after dark and large (#8-10) Irresistibles and Comparaduns tied with white or bleached hair, Catskill-style patterns, and Hi-Vis Wulffs are great options because they're easier for you to see. Use 4X tippet so the large flies don't twist your tippet and you have extra strength for landing the larger trout that feed after dark. Spinners appear at dusk, but seldom fall spent to the water until after dark. Many anglers leave too early to take advantage of the trout rising to Yellow Drake spinners, which often provide better fishing opportunities than the emerging duns.

Notes

Yellow Drakes prefer to live in warmer and slower water than Brown or Green Drakes. And since Yellow Drakes never emerge in mind-boggling numbers like the other drakes, they seldom create the same intense trout fishing opportunities. Most trout have left the warmer sections of streams by the time Yellow Drakes begin to emerge, but you may find good numbers of rising fish if the weather has been cool. If you find yourself streamside without a Yellow Drake fly pattern during a hatch, a size 10 or 12 Light Cahill usually works just as well.

Though Yellow Drake hatches appear heaviest for approximately two weeks, it's common to see the occasional Yellow Drake for up to six weeks after their initial appearance.

Pink Cahill

Female Pink Cahill dun

Para-Scud

page 191

aka *Vitreus*, Light Cahill (male), Pink Lady (female)

Epeorus vitreus, prev. *Epeorus humeralis, Iron humeralis*

Western NY: second week of June, first week of September (second generation)

Catskills: third week of June; first week of September (second generation)

Adirondacks: fourth week of June, fourth week of August (second generation)

Duration: two weeks (first generation), one week (second generation)

evening

#12–16 (8–12 mm)

Description

Nymph: (Clinger) Dark brown tails (2); body is amber on its underside with a dark olive-brown back and prominent olive plate-like gills extending from the sides; dark brown to black mottled

Above left: Mature *vitreus* nymph; *Above:* Male *vitreus* dun; *Left:* Salmon spinner (female *vitreus* spinner)

wing pads; olive-tan legs with darker heart-shaped markings on the femurs and banding on the tibia. Body, legs, and head are heavily mottled with irregularly shaped dark brown spots. **Dun:** Light gray to medium gray tails (2); pale yellowish-olive to light olive thorax and abdomen (male); olive-yellow to ginger thorax with pinkish-red to deep orange-red abdomen (female); pale, watery-gray wings; yellow-olive to pale light olive legs banded with dark brown heart-shaped markings on the femurs. **Spinner:** Light gray tails (2), ginger thorax with pinkish-red to orangish-red abdomen (female) or dark olive thorax and abdomen (male), clear wings, creamish-tan to olive-tan legs barred with darker heart-shaped markings on the femurs. Resembles *Stenacron interpunctatum*; females look like *Ephemerella invaria*; males look like *Epeorus pleuralis*.

Emergence

Epeorus vitreus begins its initial emergence when mountain laurel is blooming and oxeye daisies are just starting to bloom. Hatches are significant for about two weeks, but duns may continue to appear through July. A second generation, approximately one size smaller, hatches in September when aster is blooming. *Vitreus* hatches commence in mid to late evening, and they usually continue until dark.

Behavior and Tactics

Dead-drift dark amber to brown Hare's Ears (tied wide and flat) through riffles before the hatch begins. Nymphs are poor swimmers, so they crawl from fast currents into slow or slack water to emerge. To imitate the migrating nymphs, try stripping patterns heavily weighted with tungsten beads or extra split shot from faster currents toward slack water. Nymphs usually hold tight to submerged rocks and emerge on the stream bottom, rising to the surface as fully formed duns. Match this by swinging weighted wet flies or a wet and a nymph from fast water into slack areas.

Duns seldom spend much time on the water after emerging, so it's uncommon for large numbers of trout to feed on them. A Para-Scud is a good imitation of a dun breaking through the surface film. *E. vitreus* duns often stand erect on their legs for brief moments before they fly, a posture best imitated by stiff-hackled Catskill fly patterns. Occasionally twitching or skittering these patterns to sporadically rising trout elicits a strike.

Spinners usually fall in the evening or at dusk, but they may be delayed until early the next morning during especially warm weather. There is a distinct color difference between the males and females, and trout may become selective to either of them. Make sure to have both dark olive and pinkish-red spinners.

Notes

E. vitreus, like *E. pleuralis* (Quill Gordon), is intolerant of water pollution and is found only in New York's purest trout waters. The distinct colors of male and female duns are magnified when the duns transform into spinners. The females turn a deep pinkish-red, the color of salmon fillets, prompting some anglers to call them Salmon Spinners.

It is not widely recognized or scientifically proven that *vitreus* produces two generations per year, but based on my observations and stream samplings, I believe they do, though maybe not every year. I have observed *vitreus* in the fall some years and in other years I have not. Perhaps weather conditions or some unknown factor causes this second generation to appear only some years, similar to the way that cold water releases on the Upper Delaware cause multiple Little Sulphur generations. I believe that *vitreus* is the mythical Peach Fly that angling writers described appearing on the Beaverkill in the fall.

The female spinner photograph shows the fly extruding its eggs. Notice how the deep orange color leaves the body with the eggs. An orangish-tan spinner would best imitate this fly after all of the eggs have been laid.

Summer Blue Quill

Male Summer Blue Quill dun

aka Paralep, Blue Dun, Little Mahogany Dun
Paraleptophlebia species
Western NY: second week of June
Catskills: third week of June
Adirondacks: fourth week of June
Duration: three months
morning
#18–20 (6–8 mm)

Jim's Paralep Emerger

page 192

Summer Blue Quill nymph

Male Summer Blue Quill spinner

Description

Nymph: (Swimming) Light brown tails (3), lighter tan near the tips; dark gray to black wing pads; dark gray-brown to black body; tan legs; and gray, forked gills. **Dun:** Dark grayish-brown to pale gray tails (3), dark gray body, grayish-tan legs, and slate-gray wings. **Spinner:** Light gray tails (3); dark brown body (female) or hyaline abdomen with the last few segments iron gray (male); dark brown legs (female) or light gray legs (male); and clear wings. Male duns resemble *Ephemerella deficiens* and male spinners resemble Trico spinners.

Emergence

Summer Blue Quills hatch almost every morning throughout the summer, at the same time as milkweed is in full bloom. Hatches can be important when the weather is unusually cool or inclement and the newly emerged duns can't readily take flight. Look for the hatch to appear on a cool, inclement summer morning, often as early as 6 AM.

Behavior and Tactics

Fish dark Pheasant Tails in the slower water next to riffles and move them from the faster to slower water to imitate the nymphs migrating from riffles. Nymphs shed their shucks in the surface film and Jim's Paralep Emerger is an excellent imitation. Spinners lay their eggs and then fly away, seldom falling spent to the surface, making the spinnerfall often unimportant, although windy days combined with shaded, tight-canopied streams can put enough of the spinners on the water for the fish to eat them.

Notes

Many of New York's streams with Trico hatches also have Summer Blue Quill hatches from late June through much of September. The spinners appear in the air about the same time as Tricos. Summer Blue Quill spinners fly up and down, whereas Tricos move more horizontally.

Many people confuse the clear-bodied male Summer Blue Quill spinners with Tricos, but notice the small hind wing on the spinner in the photograph. Tricos do not have hind wings.

I've found many mature Summer Blue Quill nymphs before they emerge in the tails of pools that still have some current, just above a riffle. This is a great spot to fish their nymph imitations.

Golden Drake

Female Golden Drake dun

Anthopotamus distinctus
Western NY: late third week of June
Catskills: fourth week of June
Adirondacks: second week of July

Duration: ten days
evenings
#10–12 (13–16 mm)

Budding Emerger

page 192

Description

Nymph: (Burrower/Crawler) Amber to light brown tails (3); dark amber to reddish-brown body with darker brown markings on the back and feathery, light brown, forked gills that extend from the sides of the abdomen; medium brown to black mottled wing pads; prominent tusks that look like pinchers; and ginger legs banded dark brown. **Dun:** Pale yellow to cream tails (3) banded brownish-orange; pale yellow to deep gold abdomen, darker yellow or orange at the thorax; pale yellow to cream legs, front pair with darker orange banding at the knees and wrists; pale yellow to gold wings with faint light brown venations; orange stripe between the eyes. **Spinner:** Pale yellow to cream tails (3) banded orange; creamish-yellow to deep yellow belly with darker yellow to orange markings on the sides; darker yellow or orange thorax; translucent yellow wings with darker brown venations; creamish-yellow rear legs and mostly orange front legs. Resembles *Stenacron interpunctatum* and *Ephemera varia*.

Emergence

Golden Drakes are an interesting species. Their nymphs straddle the line between burrower and crawler, proving that it's difficult to put natural mayfly behavior into neat little boxes that fit our angling and entomology purposes. They tend to live in slower currents but they can also be found in sediment behind large rocks in swifter currents. They do not dig burrows like a true burrower, but instead they live in small cavities in silty substrate. Duns emerge in the surface film.

Golden Drake nymph

Female Golden Drake spinner

Behavior and Tactics

Amber Hare's Ears dead-drifted along silty, slack areas adjacent to current effectively imitate the nymphs. You can also catch fish by allowing the nymphs to dangle below you, just beneath the surface, to imitate an emerging Golden Drake. The transition from nymph to dun usually occurs very quickly in the warmer summer months when Golden Drakes emerge. But duns sometimes remain on the water for prolonged periods, making dry-fly fishing very productive when enough duns emerge to get the trout interested. Golden Drake parachutes or Budding Emergers catch fish at this time. Spinnerfalls are less vital to anglers, unless you want to fish at night. The heaviest Golden Drake spinnerfall that I've seen took place well after dark. A White Wulff or other very visible, hackled dry fly would be the best choice if you encounter a late-night spinnerfall.

Notes

Many of the former *Anthopotamus* species have now been reclassified into the species *Anthopotamus myops* and *A. verticis*, though *A. distinctus* and *A neglectus neglectus* are also still valid northeastern U.S. species. *A. neglectus neglectus* is a subspecies, along with *A. neglectus disjuncutus*, which is found in the southeastern United States. Most Golden Drakes that live in New York's trout water are *A. distinctus*.

I've found very small, yet fully formed, *A. distinctus* nymphs in the fall after their emergence has ended for the season. This strongly suggests that they have a two-year life cycle, though I've found no reference to this in fly-fishing literature.

Notice the outstretched posture of the front legs on the Golden Drake dun and spinner. This posture is very common with the varying *Anthopotamus* species.

Male and female Golden Drakes have similar-size eyes. This is unusual among mayflies; most males have much larger eyes than the females.

Some *Anthopotamus* duns have a middle tail that is shorter than the outer two tails. This is common in *A. distinctus* specimens that I've collected on the Delaware River's Main Stem. But I've noticed something unusual about other *Anthopotamus* species. There were hundreds of Golden Drakes one night in 2011 at a rest stop along the Susquehanna River in Pennsylvania, so I had the opportunity look closely at many of them. Some had the shortened middle tail and some did not. The flies I found were most likely *A. neglectus neglectus*, not *A. distinctus*, so perhaps the variable tail length is indicative to that species. I have been unable to find any mention of tail length variation in fly-fishing literature.

Many anglers confuse Golden Drakes (*Anthopotamus*) with Light Cahills (*S. interpunctatum*) because both mayflies are fairly large and yellow or orange colored. But all Light Cahill duns and spinners have two tails, while Golden Drakes have three.

Trico

Tricorythodes species, prev. *Tricorythus* species
Western NY: late in the first week of July
Catskills: second week of July
Adirondacks: second week of July
Duration: twelve weeks
morning
#22–26 (3–4 mm)

Female Trico Antron Wing Spinner, Sunken Male Trico, Trico CDC Comparadun

(See the appendix for all pattern recipes.)

PHOTO BY GREG HOOVER

Trico nymph (top), *Caenis* nymph (bottom)

Female Trico spinner with egg sac

Left: Caenis spinner caught in a spider's web; *Above:* Female *Caenis* dun

Description

Nymph: (Crawler) Tan to brown tails (3) that are lighter at their tips; olive-brown body with dark brown markings; dark brown, mottled wing pads; tan legs. **Dun:** Light gray tails (3); pale to medium olive body (female) or brownish-olive body (male); pale gray wings (no hind wing); light brown to cream legs. **Spinner:** Short, light gray tails (3) (female) or long, dark brown tails twice the body length (male); cream abdomen and dark brown thorax (female) or blackish-brown thorax with lighter ribbed abdomen (male); clear wings; pale cream legs. Resembles *Caenis* and summer and fall *Paraleptophlebia* species (spinners).

Emergence

Tricos begin hatching when spotted knapweed, mullein, wild onion, and wild rhododendron are blooming and continue emerging until the end of October, usually ending after the first or second frost. Male duns emerge the night before mating takes place, somewhere between 10 PM and 2 AM. Female duns emerge early in the morning—from 6 to 9 AM—depending on the air temperature. Mating, egg laying, and the spinnerfall usually occur between 7 and 10 AM, but this timing can be accelerated or delayed if the weather is unusually hot or cold. When anglers talk about fishing a Trico hatch, they are usually referring to the spinnerfall.

Behavior and Tactics

Most anglers completely ignore Trico nymphs, even though a small weighted Pheasant Tail fished through riffles and pools in tandem with a larger, more visible dry fly like an *Isonychia* parachute can be effective. Trico nymphs can hatch just about anywhere in the stream: on the bottom, at the surface, or by crawling onto submerged rocks and vegetation.

Trout feed on emerging female duns before the spinnerfall, and you can catch them with #20-24 CDC Comparaduns or thorax patterns fished on 6X or 7X tippets–but spinnerfalls provide the best fishing opportunities during the Trico hatch. Male spinners swarm about 10 to 20 feet above fast water. Females enter the swarm, mate, and then move out of the swarm (hovering 2 to 5 feet above the surface), flying upstream to lay their fertilized eggs. Mated females carry dark-olive egg sacs. It's not always necessary to fish patterns tied specifically to imitate males or females, especially during the season's earliest Trico spinnerfalls. But after a few weeks, the trout often become more selective. You may have to fish a female spinner pattern (with a white or cream body) at the beginning of the spinnerfall and then switch to a male spinner pattern (with a black body) at the end to copy the males falling to the water.

Since Tricos hatch every day for several months and there is plenty of fishing pressure, trout get selective. Use a long, 12- to 15-foot leader and 6X or finer tippet. Sometimes you may have to fish #26 rather than the standard #22 or #24. Small Trico patterns are difficult to see, so it helps to fish them in tandem with another, more visible, dry fly. You can also tie several Trico spinners to a larger hook (#16) to imitate a cluster of spinners. Some anglers set the hook whenever they see a rise near where they think their fly is. A slight lift of the rod tip is all that's needed. If you set too hard, and the fish hasn't taken your fly, you might scare it. But if the fish has taken the fly, and you set hard, you'll probably break it off.

Notes

The heaviest Trico spinnerfalls occur in sections of the stream without heavy canopy, which would prevent the normal mating flight. Even a small opening in the tree cover along a shoreline can provide enough room for Tricos to mate. Trico spinners and Summer Blue Quill spinners appear in the air 10 to 20 feet above the surface at the same time (See entry for Summer Blue Quill, page 78).

There are several species of Tricos (*Tricorythodes stygiatus, Tricorythodes minutus, Tricorythodes allectus, Tricorythodes fictus*), but you don't need to be able to tell the difference to fish them.

Neither Tricos nor *Caenis* have a hind wing. But notice the white body of the *Caenis* dun and the prominent brown spot on the top of the white *Caenis* spinner's back. Tricos are never completely white, so if you see a very tiny predominately white mayfly, it is most likely a *Caenis*. *Caenis* usually hatch and the spinners fall in the evening or nighttime, but Tricos are a morning event. You can tell Trico and *Caenis* nymphs apart by the shape of their gills. Notice the square-shaped gills of the smaller *Caenis* nymph compared to the more triangular-shaped Trico gills.

A sunken Trico, imitating spinners that have been pulled subsurface after floating into a riffle, fished beneath a larger dry fly like a caddis, can be a great aid for those who cannot see, or do not like to fish, small dry flies. A black-bodied CDC Comparadun can also work well. Because the white CDC wing stands upright, it is easier to see.

Hex

Female *Hexagenia rigida* dun

Snowshoe Emerger

page 193

aka Yellow Hex

Hexagenia rigida, Hexagenia limbata

Western NY: second week of July *(rigida)*, third week of June *(limbata)*

Catskills: third week of July *(rigida)*, late in the third week of June *(limbata)*

Adirondacks: third week of July *(rigida)*, late in the fourth week of June *(limbata)*

Duration: two weeks

evenings

#8 (15–28 mm, *rigida*)

#6–8 (15–35 mm, *limbata*)

Male *Hexagenia rigida* spinner

PHOTO BY GREG HOOVER

Female *Hexagenia limbata* dun

Description

H. rigida

Nymph: (Burrower) Tan tails (3), yellow belly, dark brown markings on back, dark brown wing pads, prominent brownish-gray feathery gills, and grayish-tan legs mottled with brown. **Dun:** Yellow tails (2) with dark brown banding and a vestigial third tail; yellow abdomen with darker brown markings across the back and an orange to peach-colored thorax; yellow wings with brown venations and brown edging, particularly on the hind wing; yellow legs; and small black eyes (female). Male has much larger compound eyes, yellow on top and dark brown to black on the bottom. **Spinner:** Tan tails (2) with dark brown banding; yellow-tan belly with dark brown markings on the back; clear wings with prominent dark brown to black venations; light brown front legs and yellow rear legs; eyes remain similar, small and black on the females and large compound eyes on the males, similarly colored to the dun stage. Resembles *Ephemera varia* and *Hexagenia limbata.*

H. limbata

Nymph: (Burrower) Yellow tails (3) and belly, dark brown markings on back, dark olive-brown wing pads, prominent olive-gray feathery gills, and ginger legs. **Dun:** Light brown tails (2) with dark brown banding and a vestigial third tail; pale yellow to tan abdomen with darker brown markings across its back; olive-gray wings with dark brown to black venations; olive-brown legs and small black eyes (female). Male has much larger compound eyes, yellow on top and dark brown to black on the bottom. **Spinner:** Brown tails (2) with dark brown banding; tan and brown belly with dark brown markings on the back; clear wings with prominent dark brown to black venations; dark brown front legs and yellow rear legs; eyes remain similar, small and black on the females and large and compound on the males, similarly colored to the dun stage. Resembles *Hexagenia rigida*.

Emergence

Nymphs burrow into muddy and sandy areas in streams or lake beds. When they leave their underground homes to feed or emerge, they become vulnerable to trout. Hex nymphs swim to the surface to molt into duns, wiggling up and down at or just beneath the surface as they try to shed their nymphal skins. This struggling movement is irresistible to trout but difficult to imitate with fly patterns.

Behavior and Tactics

Before the hatch begins, dead-drift or slowly strip large, heavily weighted yellow Hare's Ears or other Drake (Hex) nymph patterns. Try fishing a lightly weighted yellow Hare's Ear or wiggle nymph with short, quick strips, or let the fly sink and slowly lift it off the bottom to imitate emerging Hexes. Use heavy tippets and prepare for violent strikes. The duns are best imitated with large, highly visible dry flies or emergers because most of the fishing takes place at dusk and well into the night. *H. rigida* seldom appears in heavy enough numbers to cause a feeding frenzy, though fish eat them. *H. limbata* can produce amazing hatches and spinnerfalls with unbelievable numbers of insects, though scenes like this are more common in the Midwest than in New York. But there are waterways in New York that have plentiful numbers of *H. limbata*. The intense *H. limbata* spinnerfalls cause two major problems for anglers: First, this extreme biomass of food on the water often makes it difficult to get trout to notice your fly. Second, because much of the action takes place after dark, it is hard to determine whether you are casting accurately or if a fish has eaten your offering. Use oversized, hackled patterns that are easier for the fish and you to see. Cut back your tippet to 3X or 4X to prevent the larger flies from twisting your tippets and to give you the extra strength for landing the larger trout that feed after dark.

Notes

Anglers often confuse *rigida* with the Yellow Drake (*E. varia*), though *rigida* is significantly larger. But notice the fine venations throughout the hind wing of both the *rigida* dun and spinner. Yellow Drakes do not have these venations covering their hind wings. *Rigida* also has only two tails, and a tiny vestigial third tail, while *E. varia* has three tails.

Hexagenia rigida is usually found in warmwater rivers or the warmer ends of trout rivers. Because of this proclivity, it's rare to find trout eating *H. rigida* duns or spinners, but it can provide some great action if you bass fish with your fly rod. *Limbata* is more often found in trout water, though it prefers lakes and ponds or slow moving, silt-laden river pools.

Both *H. rigida* and *H. limbata* nymphs look similar to the *L. recurvata* nymph (pictured on page 53). The most noticeable differences between these nymphs are the markings on their backs and their frontal processes. *Rigida* has a pointy, almost pyramidlike, projection while *limbata* is somewhat conical and *recurvata* is more rounded. Flies tied to imitate the *recurvata* nymph work just fine should you encounter *rigida* or *limbata*.

Hexagenia limbata hatches and spinnerfalls can be so intense along the Great Lakes region that they actually appear on weather radar.

In the northeast, *H. rigida* is more commonly found in large rivers and *limbata* is most often found in lakes or ponds.

Hebe

Male Hebe dun

Yellow Spun Dun

page 193

aka Pale Evening Dun, Yellow Quill
Leucrocuta hebe, prev. *Heptagenia hebe*
Western NY: first week of August
Catskills: first week of August
Adirondacks: second week of August
Duration: ten weeks
afternoon/evening
#16–18 (6–8 mm)

Female Hebe spinner

Description

Nymphs: (Clinger) pale olive-brown tails (3); olive-brown body and legs with lighter markings; dark brown wing pads; prominent flat, rounded gills extending from the abdomen. **Dun:** Creamy-yellow to pale gray tails (2); bright yellow body and legs tinted olive with dark brown markings on the back; and pale light gray wings with dark gray venations. **Spinner:** Light gray tails (2); yellowish- to tannish-brown body; pale yellow legs; and clear wings with dark brown venations. Resembles *Leucrocuta aphrodite* and *Ephemerella dorothea dorothea.*

Emergence

Hebes begin hatching in the summer when the goldenrod first blooms and last into the fall. Their initial emergences (usually during warm weather) often begin in the evening, or near dusk, after the air and water temperatures begin to cool. In the fall, Hebes often hatch in the afternoon because temperatures are cooler.

Behavior and Tactics

Nymphs prefer to live in riffles and runs but move to areas with slow to moderate flows to emerge. They are flat, like all clinger nymphs, and share the same basic body form as the Light Cahill nymph on page 67. Before the first hatches begin, dead-drift heavily weighted olive-brown Hare's Ear Nymphs on the bottom through braided water and where riffled water begins to slow into a pool. The nymphs' swimming ability is limited, and they migrate to slower water near shore, or near in-stream obstructions, to emerge into duns just beneath the surface. Imitate the naturals by slowly stripping weighted flies toward the shore.

Once some duns begin to appear, fish an unweighted Pheasant Tail just below the surface to imitate nymphs emerging into duns. Dry-fly patterns tied with trailing shucks are generally less effective for imitating Hebe adults because the duns emerge subsurface. Spinners usually fall at dusk during the summer and often provide the best dry-fly fishing opportunities for this hatch at this time of year. In cooler weather, spinners can fall in the afternoon.

Notes

Hebes are often misidentified as Sulphurs *(E. dorothea dorothea)* or Blue-Winged Olives because they are similar in size and color. However, Hebes have only two tails, and Sulphurs have three. Also, Hebe wings are pale with heavy dark-gray venations, but Sulphurs and BWOs have solid dun-colored wings without prominent venations.

Notice the large green eyes, indicative of the male, compared to the small black eyes of the female.

A dirty-yellow-bodied Spun Dun is a good Hebe imitation when the weather is cool and rainy and the duns are having difficulty escaping the water's surface. When it's warm and dry, a hackled fly such as a Thorax Dun is a better option to imitate the way duns quickly escape from the water.

White Fly

Male White Fly dun

Ephoron leukon
Western NY: second week of August
Catskills: second week of August
Adirondacks: third week of August

Duration: one week
evening
#12–16 (12–14 mm)

Hi-Vis White Wulff

page 193

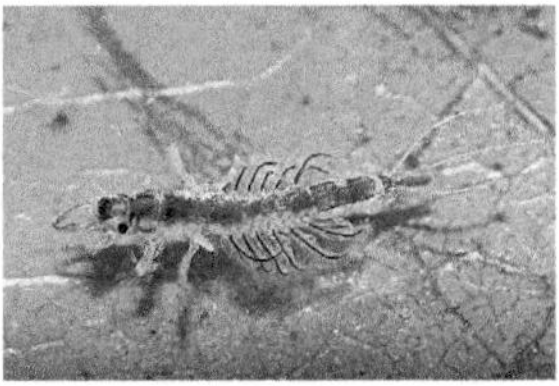

White Fly nymph

Description

Nymph: (Burrower) Pale gray tails (3), body, and wing pads; cream legs. **Dun:** Three (female) or two (male) pale gray tails; pale cream, yellowish (female), or chalky white (male) body; pale gray wings; front pair of legs is dark brown and back two pairs are white. **Spinner:** (Male only) Pale gray tails (2); chalky white body, except the last two segments, which are darker; clear wings, frosted with white; rear two pair of legs white and front pair dark brown.

Females are larger (about a size 12), and they have two tails. They also have a yellowish cast to their abdomens from the two large egg sacks they carry, changing to chalky white after they have deposited their eggs. Females have strange-looking atrophied legs that prevent them from landing. Males have two tails and are slightly smaller than the females (about one hook size). The males' legs are also weak but functional enough to hold on to the females during mating. Resembles *Maccaffertium modestum,* and *M. pulchellum.*

Emergence

White Flies usually begin hatching around the second week of August when the goldenrod has been blooming for a week or more. The hatch continues for anywhere from ten days to as long as two weeks, lasting even longer on some waters. A few of the flies may start appearing over the water's surface around 7 or 8 PM. But the hatch builds as darkness approaches, and it continues well past dark on most evenings. Occasionally, you won't see any White Flies until dark, and then suddenly, in an explosion of life, the air fills with bugs and the trout begin to feed.

Behavior and Tactics

E. leukon nymphs are sensitive to light, and they seldom leave their burrows during the day. To prepare for their emergence, the nymphs leave their burrows about an hour before dark, and you can catch fish with heavily weighted, pale gray Hare's Ears dead-drifted or slowly stripped along the stream bottom before the hatch begins. At dusk, the nymphs undulate toward the surface, pausing approximately 10 to 12 inches below it, and begin drifting as they attempt to free themselves from their nymphal shucks. Fish nymphs the same way as you would emerging Green Drake nymphs (see page 48). Perhaps the most effective method for fishing near the beginning of an evening's White Fly hatch is with a tandem. Drop a lightly weighted Hare's Ear 6 to 10 inches below a White Wulff and fish the rig at the tails of riffles and through the pools.

Once they emerge and escape from the water, the males transform into spinners and both sexes begin the search for an acceptable mate. The males fall spent to the water immediately after mating. The females lay their eggs, and then they too fall spent.

Intense White Fly hatches and spinnerfalls cause two major problems for anglers: First, there are often so many real flies on the water that it's difficult to get the trout to notice an artificial fly pattern. Second, because much of the best dry-fly action takes place well after dark, it's often difficult for anglers to see their fly patterns to know if they are casting into a trout's feeding lane or whether a fish has eaten the fly.

Oversized Hi-Vis White Wulffs (#12) are easier for you to see after dark and stand out a little better for fish. Cut back your tippet to 4X for the extra strength you need to land the larger trout that feed after dark. Trout also rise to a White Wulff skittered across the surface during the hatch/spinnerfall. When a rising trout won't take your fly, try dragging the fly just a few inches in front of the rise to induce a take.

Regardless of how many bugs you have seen that evening, if there are White Flies in the spider webs or collected in back eddies from the night before, do not leave until you are sure that the hatch won't be occurring. This means you must stay until dark. Look for rising fish in unusual places after dark. The trout's natural wariness during the day diminishes after the sun sets. Trout feed in shallow water a few inches from the bank or even within a rod's length of you.

Notes

Ephoron leukon prefers warmer sections of clean rivers and streams. Often trout have vacated the warmer sections of streams by the time the flies begin to emerge unless the weather has been cool.

Only male White Flies change from duns to spinners. A dead giveaway that you are experiencing a White Fly hatch is when you see males flying upstream, a few feet above the surface, still attached to their subimagal pellicle, which look like parachutes. Females are sexually mature adults after they transform into duns and do not molt into spinners. Entomologists debate whether males must land and rest to change into spinner, or whether they transform while they are flying, but this is unimportant for fishing.

Big Slate Drake

Catskill Style

(female *atrocaudata*), page 194

aka Hex

Hexagenia atrocaudata

Western NY: late in the second week of August

Catskills: late in the second week of August

Adirondacks: late in the third week of August

Duration: one week

dusk

#8–10, 2XL (16–24 mm)

Gray Hackle-Stacker

(male *atrocaudata*), page 194

Big Slate Drake nymph

Male Big Slate Drake dun

Female Big Slate Drake spinner

Male Big Slate Drake spinner

Description

Nymph: (Burrower) Short tan tails (3), yellowish-tan belly, darker tan or brown markings on back, dark brown wing pads, prominent brownish-gray feathery gills solid in their center, and orangish-tan to ginger legs. Frontal process is rounded, sometimes with a dimple in the center not deeply cupped like Yellow and Green Drakes but more pronounced and prominent than the Dark Green Drake. **Dun:** (Female) Olive-black tails (2) with tan banding and a vestigial third tail; yellow body, darker across its back; tan wings with brown venations and hind wing edged light brown; yellow-brown front legs with yellow rear legs; and small olive-brown compound eyes. (Male) Black tails (2) with tan banding and a vestigial third tail; tannish-olive gray body, darker across its back; tan wings with brown venations and hind wing edged dark brown; tannish-black front legs with yellow rear legs and large red compound eyes. **Spinner:** Dark gray tails (2) with tan banding; yellow to tannish-gray body; clear wings with dark brown venations and dark edging; dark olive-brown front legs and yellow rear legs. Resembles *Isonychia bicolor* (male), though *Isonychia* are much smaller.

Emergence

Hexagenia atrocaudata is fairly common, particularly in larger rivers and streams, though it seldom appears in large numbers. Hatches begin in August when goldenrod is just beginning to open, and often overlap with White Fly hatches. Duns emerge just before and after dark. The spinners begin soaring high above the water and free-falling toward the surface like glider planes around 7 PM.

Behavior and Tactics

The large nymphs are sensitive to light and generally remain tucked in their burrows during the day, leaving only at dusk or at night (earlier in overcast weather) to feed or emerge. The nymphs swim to the surface with an exaggerated undulating movement to emerge into duns, though this often occurs in shallow areas near shore where trout cannot always reach them. Imitate nymphs with large, weighted tan Hare's Ears slowly stripped along the bottom. Since it will most likely be dark, forgo the indicator and just try to feel the strike. Use heavy tippet.

The huge spinners appear over the water in the evening, well before mating, laying eggs, and falling spent to the water. Trout don't start to feed on the spent spinners until dusk or after dark, but once they have been hatching for a week or more, trout will take large Hex spinner patterns fished before it's dark, even though they won't feed on the real thing until later. After dark, huge (#4-8) Catskill-style drys dead-drifted or gently twitched in the vicinity of a trout work well. Antron-wing spinners work as well as anything to imitate the enormous spinners, but keep your tails short enough that the stiff fibers don't prevent the fish from taking the fly.

Notes

All *Hexagenia* species have three tails. Two are long and prominent, but the vestigial middle tail is short and difficult to see.

Male and female *atrocaudata* duns look significantly different, but notice how similar their coloration becomes as spinners.

Though both Green Drakes and Big Slate Drakes are burrowers, they live in different habitats. Green Drakes live in more coarse substrate throughout a streambed, but Big Slate Drakes prefer very fine silt the consistency of talcum powder, the kind of stuff that pulls your wading boot off if you step in it. This type of silt is most often found in slack areas near the shore. These areas are often so shallow that the nymphs and emerging duns are not available to trout. Spinnerfalls are by far the most important aspect of this hatch for catching fish.

H. atrocaudata's greatest predator is the water strider, which aggressively attacks the emergers. *H. atrocaudata* duns flee the surface faster than any other mayfly I've studied, most likely to avoid being eaten by the many water striders that inhabit shallow areas near the shore, but this also greatly diminishes the insect's availability to trout.

H. atrocaudata have a two-year life cycle. For much of those two years, the nymphs remain deep in the silt, unavailable to trout and even too deep for my collection attempts for this book. The nymphs were not shallow enough to be collected until two weeks before their emergence.

4 Caddisflies

Caddisflies are divided into five families, which differ according to their larval habitats. Tube-case makers live in round or rectangular cases that they construct by gluing together minute bits of plant material or sand and small rocks. Net spinners build silken nets to collect food. Beside their nets, they construct small homes from bits of sand or pebbles, where they live between feedings. Free-living caddisflies, the simplest of all caddis families, do not build homes. They generally wander beneath rocks, just like many mayfly and stonefly species. Purse-case makers live as free-living caddisflies until their final instar. After their final instar, purse-case makers construct a rounded case from plant or rocky material where they live until they hatch into adults. Most of these caddisflies are small, and anglers often refer to them as "microcaddis." Finally, saddle-case makers construct semicircular cases, resembling shells, out of larger pebbles.

The caddisfly's life cycle is complete. Whereas a mayfly changes directly from a nymph into a winged, subadult (called a subimago), the caddisfly's life cycle has an extra step. A caddisfly hatches from an egg and becomes a larva. The larva feeds and grows throughout the year, usually molting five times. At approximately the same time each year, depending upon the species, the larva begins its transformation to adult on the stream bottom, inside a pupal shuck. This metamorphosis from larva to pupa (called a resting stage) lasts from several days to a few weeks, again depending upon the species. Once the pupa is fully formed, it chews through its shuck and crawls, drifts, or swims toward the surface, shedding the shuck in the process. The pupa eventually breaks the water's surface film as a fully formed, sexually mature adult. The adult flies to streamside vegetation where it prepares to mate.

Male and female caddisflies gather over riffles to mate while they're flying. After mating, the males die; the females deposit their eggs, and then they also die. Some females drop their eggs on the water's surface by fluttering and dipping their egg sac until it falls free. Others swim or crawl to the stream bottom to distribute their eggs. This life cycle, from egg to mating and laying eggs, is usually completed in one year. But some species, like *Psilotreta labida*, have a two-year life cycle.

CADDIS LIFE CYCLE

Pupa

Adult

Egg layer

Standard caddis larva imitations

Olive Caddis Larva, BDHD Walt's Worm, Bird of Prey Caddis, Walt's Worm, Green BDHD Caddis Larva, Scuddadis, Black BDHD Orange Caddis larva (See appendix for recipes.)

Wet flies

Partridge and herl, Quill Gordon Style Wet, Partridge and Green, Traditional Wet. (See appendix for recipes.)

CADDISFLY HATCH CHART

Scientific name	Common name/size	Time of day	Hatch begin in Western New York
Dolophilodes distinctus	Winter Caddis #18-20	midmorning (winter); early morning (summer)	second week c February (winter); seco week of July (summer)
Chimarra species	Little Black Caddis #16-18	late morning, afternoon, evening	third week of March
Brachycentrus numerosus and *fuliginosus*	Grannom #12-16	morning and afternoon	fourth week of April
Brachycentrus appalachia	Apple Caddis #14-16	morning and afternoon	first week of May
*Diplectrona modesta**	Dark Brown Caddis #12	morning and afternoon	second week o May
Various *Hydropsyche* and *Symphitopsyche* species	Spotted Sedge #14-18	morning and afternoon	second week o May
Cheumatopsyche (various species)	Little Sister Sedge #16-20	morning and afternoon	second week of May
Psilotreta labida and *frontalis*	Dark Blue Sedge #12-14	late afternoon and evening	third week of May
Rhyacophila lobifera (and related species)	Green Caddis #12-16	morning and evening	fourth week of May
Pycnopsyche scabripennis	October Caddis #10	evening	fourth week of September

*Minor hatches

Hatch begins in Catskill region	Hatch begins in Adirondack region	Hatch duration
urth week of ebruary (winter); iird week of July ummer)	first week of March (winter); fourth week of July (summer)	three months, but heaviest for two weeks (winter generation); two months (summer generation)
rst week of April	third week of April	seven months
rst week of May	second week of May	two weeks
econd week of May	third week of May	one week
hird week of May	fourth week of May	six weeks
hird week of May	fourth week of May	three months
hird week of May	first week of June	one month
ourth week of May	first week of June	two weeks
irst week of June	second week of June	two months
irst week of October	first week of October	three weeks

Winter Caddis

Male Winter Caddis adult

Arty's Black Caddis

page 196

Dolophilodes distinctus

Western NY: second week of February (winter); second week of July (summer)

Catskills: fourth week of February (winter); third week of July (summer)

Adirondacks: first week of March (winter); fourth week of July (summer)

Duration: three months, but heaviest for two weeks (winter); two months (summer)

midmorning (winter); early morning (summer)

#18–20 (5–8 mm)

Description

Larva: (Net Spinner) Pale creamy-yellow body with darker orange head (looks nearly identical to the *Chimarra* larva on page 112 but with a much lighter body color). **Pupa/Adult:** Dark brown to black body; dark brownish-black to dark gray wings mottled with light brown dots (winter generation female adults have no wings); light brown to brown legs. Resembles several *Brachycentrus* and *Chimarra* species.

Female Winter Caddis adult

Winter Caddis larva

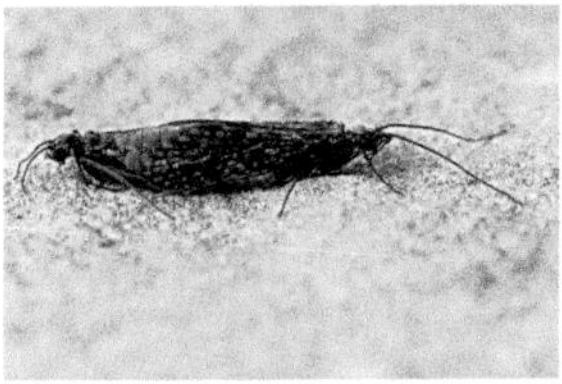
Mating Winter Caddis

Underside of Winter Caddis

Emergence

Though they are often called Winter Caddis, *Dolophilodes distinctus* has two emergences: one in winter and one in summer. There is little phenology for winter-hatching species, but Thomas Ames, in his book *Caddisflies*, states that *Dolophilodes* winter emergence begins as early as November and peaks in February when the early *Baetis* begin hatching. The summer emergence continues from July through August.

The highest concentration of *Dolophilodes* that I have found, and the most widespread reports of fish feeding on them, have been in late February.

Behavior and Tactics

Dead drifting light yellow bead-head caddis larva through riffles catches fish. Pupae swim to the surface and then skitter across the water to the shore to emerge. It's difficult to imitate the pupa running across the surface, but a Griffith's Gnat, tied with black hackle, can be skittered across the water to imitate this behavior. You will have difficulty seeing this fly and will most likely have to feel the strike. In winter, the winged males are often seen lining streamside rocks, peering into the water, as if they are waiting for the wingless females to crawl out of the water. Fish rise to winged male fly patterns, and Arty's Black Caddis is a good imitation, though standard black Elk Hair Caddis, tied with a gray wing, also works.

Notes

Though the male's wings appear obviously mottled brown and black in the photographs, their wings often appear light gray when they are first captured streamside.

The mating *Dolophilodes* photograph may be fairly rare. The bugs began mating in the romantic confines of a plastic fly cup in my refrigerator after being captured by Patrick Williams. I use the refrigerator's cold to slow the insect activity so I can photograph them. But these two *Dolophilodes*, already accustomed to the cold by hatching in the winter, didn't seem to mind.

Dolophilodes distinctus is most important during its winter emergence because there are few feeding options for the fish, and they often key on this small caddis. During the summer, *D. distinctus* hatches are overshadowed, and often overlooked by anglers, due to the wide variety of insects and other caddis species hatching simultaneously.

Only male *Dolophilodes distinctus* adults have wings in the winter. But both male and female adults have wings in the summer. I believe that there is a natural reason for this. If winter females had wings, they would most likely fly away from the water, leaving them to freeze as the air temperature drops at night. But because they can only crawl, they stay nearer to the warmer microenvironment of the water's edge, allowing them to survive the cold. The males are seeking to reproduce, so they also stay near the warmer water, looking for the females.

The *Dolophilodes* larva photographed here appears shorter than it really is. The insect is curling its abdomen in a defensive posture.

Little Black Caddis

Little Black Caddis adult

CDC Caddis

page 196

Chimarra species
Western NY: third week of March
Catskills: first week of April
Adirondacks: third week of April
Duration: seven months (combination of all *Chimarra* species emergences)
late morning and afternoon (spring and fall), evening (summer)
#16–18 (6–8 mm)

Description

Larva: (Net Spinner) Amber to orange body with darker orange head. **Pupa/Adult:** Dark brown to black body; black to dark gray wings; dark brown to black legs. Resembles several *Brachycentrus* species and *Psilotreta labida*.

Little Black Caddis larva

Mating Little Black Caddis

Emergence

Chimarra first emerge when forsythia start to bloom, beginning each day around noon and continuing through the afternoon. Emergences trickle through the summer and become heavier again during fall.

Behavior and Tactics

Chimarra spend most of their lives beneath rocks, feeding on small particles trapped in their nets. But they travel to the tops of their rocky homes and crawl, or swim, to shallow areas near shore to emerge. Dead-drift standard orange bead-heads and Bird of Prey larva patterns along the river bottom or slowly strip them toward shore before and during an emergence. Adults, when they first emerge, are generally less important to anglers because trout may not be holding in shallow areas near shore where the adults emerge at this time of year. But egg-laying females can provide fantastic dry-fly opportunities when they fall helplessly to the water after depositing their eggs, providing an easy meal for trout. These spent caddisflies are best imitated by dead-drifting CDC Scud Caddis patterns to sipping fish.

Notes

Chimarra caddis live throughout New York, usually in medium to large creeks and rivers. Their hatches sometimes overlap with *Baetis*, Blue Quill, and Hendricksons. Their hatches are often overshadowed by Hendricksons, frustrating anglers casting mayfly patterns to trout selectively feeding on caddis.

Chimarra caddisflies are sometimes confused with adult Alderflies (see photo on page 160). But the two unrelated insects look quite different when seen together.

The *Chimarra* genus includes several species that hatch at different times. They are most important in the spring and fall, though an orange caddis larva tied to imitate *Chimarra* can catch fish all year long.

Chimarra pupae emerge into adults by climbing out of the water onto dry land. Their empty cases can often be found on top of streamside cobbles.

Grannom

Grannom adult

Peacock Caddis

page 196

Brachycentrus numerosus and *B. fuliginosus*
Western NY: fourth week of April
Catskills: first week of May
Adirondacks: second week of May
Duration: two weeks
morning and afternoon
#12–16 (7–13 mm)

Above left: Grannom larva in case; *Above:* Underside of Grannom; *Left:* Grannom egg-layer

Description

Larva: (Tube-Case Maker) Brown head and legs, bright green body. **Pupa/Adult:** Greenish-black body, mottled brown wings, greenish-black legs. Resembles *Chimarra atterima* and *Psilotreta labida.*

Emergence

Grannoms emerge when marsh marigolds and forsythia are in bloom. They hatch late in the morning to midday and continue through the afternoon. They can coincide with the end of the Hendricksons and the beginning of the March Browns.

Behavior and Tactics

Larvae live in well-oxygenated areas of current, inside dark brown tubes created from bits of plant matter and shaped like flat-sided ice cream cones or oil derricks. The pupae swim to the surface film to emerge into adults. Trout sometimes ignore the newly hatched adults on the surface, opting to feed on the subsurface larvae and pupae. Dead-drift green bead heads and Bird of Prey larva imitations or swing wet flies through riffles.

Egg-laying females provide an easy meal for trout when the river bottom isn't full of larvae and pupae. The females fall helplessly to the river's surface as they try to rid themselves of their egg balls. Imitate this behavior by either dead-drifting or gently twitching Peacock, Elk Hair, and Henryville dry-fly caddis patterns.

Notes

Grannoms are closely related to the Apple Caddis (they belong to the same genus), though Grannoms are darkly colored and much more widespread in New York than the light-colored Apple Caddis. Hatches often overlap with Apple Caddis emergences in streams where both species are present.

Though the Peacock caddis may not have been designed to imitate the Grannom, its peacock herl body represents Grannom body coloration very well.

Apple Caddis

Apple Caddis

aka Shad Fly
Brachycentrus appalachia
Western NY: first week of May
Catskills: second week of May
Adirondacks: third week of May
Duration: one week
morning and afternoon
#14–16 (11–13 mm)

MT's Apple Caddis (John Miller)

page 196

Apple Cadddis case

Description

Larva: (Tube-Case Maker) Brown head and legs, bright green body. **Pupa/Adult:** Apple-green abdomen, medium ginger thorax and legs, cream to light tan wings that are lighter at their tips. Resembles *Rhyacophila* species.

Emergence

Apple Caddis usually live in medium to large creeks and rivers. Their emergence begins late in the morning to midday and continues through the afternoon. They appear when apple trees are heavily budded and in bloom and Black Quills are hatching.

Behavior and Tactics

Larvae live in well-oxygenated areas of current, inside brown tubes created from bits of plant matter and shaped like flat-sided ice cream cones. Pupae swim to the surface film to emerge into adults. Dead-drift green bead-head larvae and Bird of Prey larvae imitations or swing wet flies through the riffles to imitate the larvae and pupae.

Trout sometimes ignore the thousands of newly hatched Apple Caddis adults on the surface, opting to feed subsurface on the larvae and pupae. Egg-laying Apple Caddisflies, near the end of their yearly emergence, provide an easy meal for trout when the river bottom isn't full of larvae and pupae. The females often fall helplessly to the river's surface as they try to rid themselves of their egg balls. Imitate this behavior by either dead drifting or gently twitching CDC, Trude-style, Elk Hair, and Spun Head Deer Hair Caddis patterns.

Notes

Though not present in as many New York trout streams as Grannom, the Apple Caddis can be important in the watersheds where it lives. The species is similar to Grannoms, differing only in wing and body coloration. The Apple Caddis's abdomen is brightly colored like a green apple, whereas the Grannom is nearly black. The two hatches often overlap, and sometimes coincide with the end of the Hendricksons and beginning of the March Brown mayflies.

In New York's Upper Delaware River system, Apple Caddis often emerge during the American shad migration to the trout sections of the river, earning them the local name "Shad Fly." The use of this colloquialism creates a good argument for the importance of using scientific nomenclature when discussing aquatic insect hatches. In much of central Pennsylvania, including the Penns Creek watershed, the term "Shad Fly" is used for the Green Drake dun (*Ephemera guttalata*).

Notice that the Apple Caddis's case is nearly identical to the Grannom's, illustrating their shared genus. Some anglers also call the Apple Caddis "Popcorn Caddis."

Spotted Sedge

Spotted Sedge adult

Bivisible (the Dette shop)

page 197

aka Tan Caddis
Hydropsyche species and
Symphitopsyche species
Western NY: second week of May
Catskills: third week of May
Adirondacks: fourth week of May
Duration: three months
morning and afternoon
#14–18 (8–15 mm)

Description

Larva: (Net Spinner) Dark brown to black head and upper back, olive-brown to tannish-olive belly and lower back. **Pupa/Adult:** Creamish-tan to light brown body and legs and light brown wings. Resembles *Cheumatopsyche* (but *Hydropsyche* and *Symphitopsyche slossonae* are larger and more bland in color).

Spotted Sedge larva

Spotted Sedge nets

Emergence

These species are some of the most important and prolific caddisflies in New York. Collectively, they form a long-lasting group of insects which begin hatching in mid to late spring when dame's rocket is just starting to bloom and continue throughout the summer and into the early fall. Early hatches often occur with the March Brown, though because of the long emergence period of many species, hatches overlap with those of many other aquatic insects.

Behavior and Tactics

Larvae prefer to live in riffles and runs, anchored to submerged rocks with a silk line. They are difficult to imitate with artificial flies because of the way in which they dangle in the current. Gary LaFontaine, in his book *Caddisflies,* suggested several ideas for imitating this behavior, including coloring tippets white to suggest the anchoring line. But dead-drifted olive bead-heads and Bird of Prey larvae imitations work well to imitate the naturals drifting with the current before swimming toward the surface to molt into adults. Adults often remain on the water's surface for extended periods as they try to recover from molting. Adult females dive to the river bottom to lay their eggs and then float to the surface, drifting spent like a spinner in the film. Dry-fly patterns work well when the caddis first molt into adults and when they return to lay eggs. Dead-drift X Caddis, Bivisibles (especially in riffles), and CDC Caddis to imitate the adults and spent egg-layers. Swing wet-fly patterns to imitate the diving egg-layers, or dead-drift weighted dry-fly patterns to imitate adults that have finished laying their eggs.

Notes

The Tan Caddis is more tolerant of certain types of pollutants, such as sewage effluent and farming runoff, than most aquatic insect species, which allows them to thrive in many New York trout streams where other aquatic species cannot.

Weather can dictate the timing of this hatch (and most other caddis hatches as well). If the air and water temperatures are unusually warm, the hatch may not begin until early evening, or it may wait till morning. During cloudy, cool weather, the hatch sometimes occurs in the afternoon.

Some Tan Caddis actually have an olive sheen on their abdomens as adults, confusing anglers who may think they are Green Caddis.

Though the Bivisible doesn't really look like a caddis, its bushy construction makes it a good, high-floating choice for matching Tan Caddis hatches in riffles.

Little Sister Sedge

Little Sister Sedge adult

aka Olive Caddis
Cheumatopsyche spp.
Western NY: second week of May
Catskills: third week of May
Adirondacks: first week of June
Duration: one month
morning and afternoon
#16–20 (7–12 mm)

Olive Chuck Caddis (the Dette shop)

page 197

Description

Larva: (Net Spinner) Dark brown to black head and upper back, olive-brown to tannish-olive belly and lower back. **Pupa/Adult:** Emerald green and black body, brown legs, and dark brown wings with light brown or tan spots, often with one tan spot in the wing's center. Resembles *Chimarra* species.

Underside of Little Sister Sedge

Emergence

This important group of insects hatches when lilacs are in bloom. But they are often overlooked by anglers who are concentrating on the famous mayflies that also hatch during this period, such as March Browns. *Cheumatopsyche* species are similar to *Hydropsyche* species and *Symphitopsyche slossonae,* but can live in warmer and even more nutrient-rich sections of streams and rivers. Hatches can occur in the morning or evening.

Behavior and Tactics

This species received its common name "Little Sister Sedge" because it shares many physical and hatching characteristics with *Hydropsyche* and *Symphitopsyche slossanae* larvae. The same fishing techniques for *Hydropsyche* and *Symphitopsyche slossanae* species apply for *Cheumatopsyche* species. The only significant differences between these two types of caddisflies are their size and their body and wing color as adults, and the timing of *Cheumatopsyche* hatches, which usually end by mid-June.

Notes

Their small size combined with overlapping emergences with many famous mayfly species make this caddis easy for anglers to miss. But trout love these little caddisflies. It's interesting to note that the Little Sister Sedge is called that because of its similarities with the Spotted Sedge (aka Tan Caddis). Though the two species don't look similar as adults, their larvae look nearly identical.

Dark Blue Sedge

Psilotreta labida, P. frontalis
Western NY: third week of May
Catskills: fourth week of May
Adirondacks: first week of June
Duration: two weeks
late afternoon and evening
#12–14 (12–15 mm)

Spun Head Deer Hair Caddis

page 197

Description

Larva: (Tube-Case Makers) Gray head, green body. **Pupa/Adult:** Dark bluish-gray body, occasionally with olive reflections; dark bluish-gray wings; dark gray legs. Resembles *Chimarra* species, but *Psilotreta labida* is much larger and tinted with emerald-green reflections.

Emergence

Psilotreta labida usually emerges in late afternoon and evening when oxeye daisies are just starting to bloom. This hatch coincides

Dark Blue Sedge cases

Underside of Dark Blue Sedge

with Green Drake and Sulphur hatches and often appears with the Slate Drake.

Behavior and Tactics

Larvae have a two-year life cycle and live in small tube-shaped structures, which they construct from bits of sand and gravel. Dozens migrate to shared rocks before either crawling or swimming to the surface to emerge. Look for rocks in the streambed that contain masses of *Psilotreta* cases to determine the best locations to fish the hatch. The pupae often wiggle and twitch on or just below the surface, and trout often find this movement irresistible. Imitate this behavior by twitching dry caddis patterns about one foot in front of rising fish or dangle wet and dry flies downstream in the current. Allow the flies to swing below you, but do not strip them back immediately. The current holds them near the surface as the line tightens, and the flies undulate just beneath and in the surface film. Skitter an Elk Hair Caddis across the water's surface to imitate the adult females dipping and fluttering on the surface when they return to lay their eggs.

Notes

Their dark coloration (sometimes camouflaging them on the water) combined with simultaneous emergences of mayflies make hatches of Dark Blue Sedge difficult to detect. Anglers often misdiagnose trout rising to the caddis and cast mayfly patterns to unsympathetic rising fish.

Cases from other caddis species are intermixed with the *Psilotreta* cases in the photo.

Green Caddis

Green Caddis adult

Rhyacophila lobifera and related species
Western NY: fourth week of May
Catskills: first week of June
Adirondacks: second week of June
Duration: two months
morning and evening
#12–16 (8–18 mm)

Delaware Adams (Mary Dette)

page 197

Description

Larva: (Free Living) Brown head with black markings, bright green to pale olive body. Some species have black banding along the body **Pupa/Adult:** Medium olive to green body, dark brown wings with lighter brown spots, medium olive to green legs.

Emergence

Green Caddis hatches usually begin in late morning or early afternoon and last until dark. Summer hatches often occur earlier in

Green Caddis larva

Green Caddis larva variation (different species)

Pupating Green Caddis

Underside of Green Caddis

the morning and later in the evening. Hatches usually begin when redbud starts to bloom, lilacs are getting ready to bloom, and Green Drakes have ended.

Behavior and Tactics

Larvae live in riffles and other swift-flowing sections of water and roam beneath rocks, anchored to the river bottom by a light brown silk line. To imitate those that sometimes lose their hold and are swept into the fast currents, dead-drift bright green bead heads and Bird of Prey larvae imitations through riffles. Pupae are strong swimmers and trout feed on them as they drift and swim toward the surface, preparing to emerge. Imitate them by swinging wet flies.

Delaware Adams, Henryville, X Caddis, and Elk Hair Caddis imitations fished with a dead drift in riffles or skittered accurately imitate the adults, which hatch in great numbers. Females dive or crawl to the stream bottom to lay eggs and then slowly drift spent to the surface. Imitate these submerged adults by adding split shot or beads to standard dry flies fished dead-drift.

Notes

Green Caddis hatches, due to their lengthy duration, can overlap with many mayfly and caddis species. But they are most important to fishermen when they hatch in streams with limited aquatic insect life. The Green Caddis is fairly common in all regions of New York, though it is found intermittently from stream to stream.

The Delaware Adams was created by Walt Dette for angling legend Art Lee. Though it's not really a caddis pattern (it has upright wings), its bushy profile and bright green body make it an excellent pattern to fish during Green Caddis emergences.

October Caddis

CDC Palmered Elk Hair Caddis

page 197

aka Great Brown Autumn Sedge
Pycnopsyche species, prev. *Stenophylax* species
Western NY: fourth week of September
Catskills: first week of October
Adirondacks: first week of October
Duration: three weeks
evening
#10 (19–22 mm)

October Caddis case

October Caddis larva removed from case

Underside of October Caddis

Strawman Caddis adult

Description

Larva: (Tube-Case Makers) Gray-brown head and upper back, ginger to creamy-tan body. **Pupa/Adult:** Tannish-orange body, legs, and wings. Wings have prominent black blotches. Resembles *Hydropsyche* and *Symphitopsyche* but is much larger and a deeper orange color.

Emergence

Adults hatch after the goldenrod has been open for a couple of weeks. October Caddis hatches sometimes overlap with sporadic mayfly emergences of *Isonychia bicolor, Acentrella turbida,* and *Leucrocuta hebe.* Though October Caddis emerge most often at night, they can also hatch (in smaller numbers) in the evening and morning.

Behavior and Tactics

Larvae eat decomposing leaves from trees and construct large cases from pieces of sticks and other plant debris. They live in slow-moving, placid sections of streams where leafy material collects on the stream bottom. Because October Caddisflies often emerge right before or after dark, many anglers miss seeing the naturals on the water. But they are often seen fluttering at lights outside streamside cabins or homes. Swinging October Caddis Bird of Prey larvae patterns through riffles in the morning or evening sometimes catches fish, so it's worth trying if the fishing has been slow with other techniques. However, it's generally more effective to blind-cast large, orange Elk Hair Caddis in riffles—dead-drifting, twitching, and skittering the flies. It takes a bit of faith to fish October Caddis patterns, because few (or sometimes no) naturals are visible during the day.

Notes

October Caddis live in cold streams and rivers, usually in forested areas. They provide a good-size meal when few other aquatic insects are readily available, making them important for fishermen, though the insects seldom hatch in large numbers.

The Strawman Caddis (*Limnephilus* spp.) shows a strong resemblance in both color and size to the October Caddis. Look for blotches on the wings to determine which one you're seeing: The October Caddis has large, prominent blotches on its wings and the Strawman does not have any at all.

5 Stoneflies

Stonefly nymphs hatch from an egg and spend one to three years maturing, depending upon the species, before emerging, mating, and laying eggs. They have an incomplete life cycle (like mayflies) and don't pupate. Stonefly nymphs freely wander beneath rocks in search of food. At approximately the same time each year (the exact date depends upon the species), stonefly nymphs crawl to exposed rocks or vegetation to molt into adults. Some species also emerge subsurface if conditions prevent them from reaching dry areas. The adults then fly or crawl to streamside vegetation to prepare to mate.

Stoneflies mate on the ground, not in flight like mayflies. They are generally poor flyers, and that may contribute to their land-bound mating practices. Stoneflies deposit their eggs in riffles, beginning the process during periods of low light, early in the morning or at night. Mating stoneflies do not fall spent to the water like mayflies and caddisflies, so they do not offer the trout an opportunity to eat them once mating is completed.

STONEFLY LIFE CYCLE

General stonefly nymphs

Simple Black Stone, Lee Hartman's Golden Stone, Brian Wolfkiel's Woven Stone, Rob Lewis' Realistic Stone, Kyle's Golden Stone (See appendix for recipes.)

STONEFLY HATCH CHART

Scientific name	Common name/size	Time of day	Hatch begins in Western New York
Allocapnia and *Capnia* species	Little Black Stonefly #18-22	midafternoon	second week of February
Taeniopteryx nivalis	Early Black Stonefly #12-16	early afternoon	third week of February
Strophopteryx fasciata	Early Brown Stonefly #14-16	early afternoon	first week of March
Acroneuria, Agnetina, and *Paragnetina* species	Light or Golden Stonefly #6-12	afternoon through evening or early morning	second week of May
Pteronarcys species	Giant Stonefly #4-8	afternoon through evening or early morning	third week of May (Sporadic adults may appear much earlier.)
Peltoperlids: *Peltoperla* and *Tallaperla* species	Roachlike Stonefly #14-16	midday through afternoon	first week of June
*Alloperla imbecilla**	Little Green Stonefly #14-16	afternoon and evening	first week of June
Isoperla bilineata	Yellow Sally #14-16	afternoon and evening	first week of June

*Minor hatches

Hatch begins in Catskill region	Hatch begins in Adirondack region	Hatch duration
ırd week of February	first week of March	one month
ırth week of bruary	second week of March	one month
cond week of March	fourth week of March	three weeks
ırd week of May	second week of June	six weeks
ırth week of May poradic adults may pear much earlier.)	second week of June (Sporadic adults may appear much earlier.)	one week
cond week of June	third week of June	three weeks
cond week of June	third week of June	two months
cond week of June	third week of June	two months

Winter Stonefly

Dette Stonefly Nymph (the Dette shop)

page 199

aka Slender Winter Stonefly, Slender Snowfly
Allocapnia and *Capnia* species
Western NY: second week of February
Catskills: third week of February
Adirondacks: first week of March
Duration: one month
midafternoon
#18–22 (3–8 mm)

Description

Nymph: Dirty yellow and brown legs, abdomen, and thorax; black wing cases and tails mottled with brown. **Adult:** Black tails, body, and legs; light dun wings with black venations. Wings are half-formed stubs on some adult species. Resembles *Taeniopteryx* spp.

Emergence

Some Winter Stoneflies have partially formed wings, so they cannot fly. They are smaller—size 18 to 22—and often crawl on top of snow

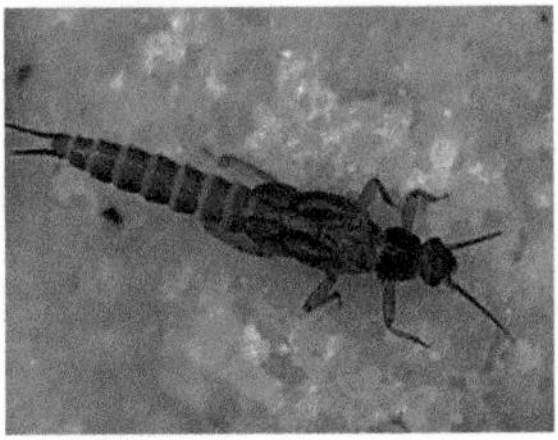

Capniid nymph

Slender Winter Stonefly adult

and ice near a stream, not on the water's surface. Nymphs migrate to the frozen edges of a stream to emerge, outside the range of most trout, which are usually holding in deeper water at this time. Some Winter Stonefly species are slightly larger—size 16 to 20—and they do have wings, but I have never witnessed one of them flying. They too seem to crawl to emerge and complete their life cycle.

Behavior and Tactics

Because *Capnia* and *Allocapnia* stoneflies cannot, or do not often, fly, adults are seldom found on the water, making dry-fly patterns unnecessary. But trout are fond of their nymphs. Any small stonefly nymph pattern can be effective, but the Dette Stonefly Nymph works particularly well. Fish the nymphs with a dead drift under an indicator along slack currents adjacent to riffles and watch closely for subtle takes. The water is still usually very cold when these flies emerge, and a trout's diminished metabolism at this time of year seldom allows for aggressive strikes.

Notes

Allocapnia and *Capnia* stoneflies are two of the first trout stream insects to emerge each year. They are a harbinger of warmer weather and better fishing to come, though they can sometimes provide good nymphing opportunities. The flies are easiest to see when there is snow on the ground. When my wife and I lived along New York's Beaverkill, I would often find Winter Snowflies near our house. This is worth mentioning because we lived 75 yards from the water, up a steep cliff. I have no idea why these tiny flies wandered so far from the river.

An easy way to differentiate between Winter Stonefly adults and Little Black Stonefly adults is their tails. You can easily see the tails of Winter Stoneflies protruding from beneath their wings on the insects' backs. You cannot see protruding tails when you are looking at the backs of Little Black Stoneflies (*T. nivalis*).

Little Black Stonefly

Black Stimulator (the Dette Shop)

page 199

aka Black Willowfly
Taeniopteryx nivalis
Western NY: third week of February
Catskills: fourth week of February
Adirondacks: second week of March
Duration: one month
early afternoon
#12–16 (9–14 mm)

Little Black Stonefly nymph

Description

Nymph: Black tails, body, and legs with faint brown stripe beginning behind the head and extending the length of the back; black wing cases mottled with brown. **Adult:** Black to dark brown tails, body, and legs; light dun wings with black venations. Resembles *Allocapnia* spp., *Capnia* spp., and *Strophopteryx faciata.*

Emergence

Taeniopteryx nivalis usually emerges in cold rivers and streams from late morning through midafternoon, and hatches are best on warm, sunny days. Snowdrops, one of the first flowers of spring, and trailing arbutus are in bloom.

Behavior and Tactics

Trout often eat adults on the water's surface because some of them hatch in the river, despite the common belief that all stonefly species crawl to riverside rocks to emerge. Though adults are on the surface of many New York trout streams, anglers will usually fare better using nymphs because the cold water reduces the trout's metabolism, making them less inclined to chase their meals on the surface. Dead-drift Copper Johns and Pheasant Tails slow and deep to interest sluggish fish.

Stonefly adults buzz and skitter across the surface, behavior effectively imitated by skittering a small, black Elk Hair Caddis or Stimulator across the surface. A late winter or early spring day with warm sunshine and a steady breeze is a great time to find trout rising to the adults. When the sun warms the water, the insects and the trout become more active, and the breeze blows more adults onto the water. In cold water, trout are more sluggish and a Flat-Water Stonefly is a better option for rising fish.

Notes

Taeniopteryx nivalis is one of the most common stonefly species in New York and seems to be more pollution tolerant than most other stoneflies.

The white, ricelike appendages extending from behind the legs are actually the Little Black Stonefly's gills.

Little Brown Stonefly

Little Brown Stonefly

aka Brown Willowfly
Strophopteryx fasciata
Western NY: first week of March
Catskills: second week of March
Adirondacks: fourth week of March
Duration: three weeks
early afternoon
#14–16 (9–11 mm)

Flat-Water Stonefly

page 199

Little Brown Stonefly nymph

PHOTO BY GREG HOOVER

Description

Nymph: Brown tails and legs; black wing cases mottled with brown. Body is olive-brown on the belly and dark brown to black on the back. Abdomen is prominently ribbed lighter. **Adult:** Brown tails, body, and legs mottled with black patches; light dun to tan wings with brown venations. Resembles *Taeniopteryx nivalis*.

Emergence

Little Brown Stonefly hatches often overlap with *Taeniopteryx nivalis* hatches in rivers and creeks where both insects are present. At this time, the forsythia is heavily budded and the white-flowered rue anemone is fully open. Just like *Taeniopteryx nivalis, Strophopteryx fasciata* emerges from late morning through midafternoon (the warmest part of the day). A warm, sunny day creates good hatching conditions for this insect and usually increases the hatch's intensity and duration.

Behavior and Tactics

Nymphs crawl onto exposed rocks, snow, or ice to emerge. They are less likely to emerge in the river than *Taeniopteryx nivalis,* but they will occasionally when high water prevents them from finding exposed rocks. Dead-drift Copper Johns or brown Hare's Ear Nymphs along the bottom or slowly retrieve them toward the shore. Although adults can drift on the surface, you'll probably catch more fish if you use nymphs (just like with *Taeniopteryx nivalis*).

When fish are rising to adults on the surface, skitter a small brown Elk Hair Caddis or dead-drift a Slow-Water Stonefly to catch them. A late winter or early spring day with warm sunshine and a steady breeze increases your chances of finding trout rising to the adults.

Notes

Strophopteryx fasciata lives throughout New York. It seems to require cleaner water than *Taeniopteryx nivalis,* so just because one of these stonefly species is present in a stream doesn't mean that both will be found.

Golden Stonefly

Golden Stonefly dry flies

Jake Villwock's Foam Egg Laying Parachute Stone, Stimulator, Jake Villwock's Foam Golden Stone (See appendix for recipes.)

aka Light Stonefly, Brown Stonefly, Yellow Stonefly
Acroneuria, Agnetina, and *Paragnetina* species
Western NY: second week of May
Catskills: third week of May
Adirondacks: second week of June
Duration: six weeks
afternoon through evening or early morning
#6–12 (14–27 mm)

Top: Golden Stonefly nymph; *Top right:* Golden Stonefly nymph (species variation); *Right and below:* Golden Stonefly adults (species variations)

Description

Nymph: Yellow to golden-brown tails; bright gold to yellow to brown body, occasionally with olive and brown reflections; black markings on the wing cases and back very prominent in some species; brown to black wing cases mottled with yellow; yellow to golden-brown legs. **Adult:** Yellow tails, body, and legs; some species have areas of black or brown markings on the underside and back; light dun wings (sometimes shaded yellow) with brown to black venations. Anglers sometimes confuse Golden Stoneflies with Giant Stoneflies (*Pteronarcys dorsata*) because both insects can be large and vary in color.

Emergence

Golden Stonefly species comprise a large group of insects with a long hatching duration, lasting from late spring through early to mid summer. Adults often first appear at dusk and coincide with the Green Drake hatch. Shortly after, dame's rocket begins blooming.

Behavior and Tactics

Most Golden Stoneflies have a two- to three-year life cycle, depending on the species. They crawl to streamside rocks (and even concrete bridge piers) and vegetation to emerge, generally during low light at, or after, sunset, though they also emerge during the day. Dead-drift Rubber Leg Beadhead Stonefly Nymphs during high to moderate stream flows in the winter and early spring or slowly strip them toward the stream banks during an emergence. Adults are clumsy fliers and often fall onto the water's surface, creating opportunities to blind-cast large dry flies. Golden Stones provide a substantial meal to trout, and fish continue to take imitations (Madam Xs or Stimulators) throughout the summer and fall, well after the hatch has ended for the season. For this reason, a Golden Stonefly pattern with a nymph dropper is a great choice for blind casting during the summer.

Notes

Golden Stoneflies hatch throughout New York, but they generally live in swift-flowing riffles and runs of clean, cold streams adjacent to forests.

Anglers lump many species of gold-, yellow-, and brown-colored stonefly adults together under the common name Golden Stoneflies. And though Golden Stonefly nymphs can look very different physically (see the photographs above), techniques for fishing the different species are identical. But it is wise to sample the stream you're fishing to see which Golden Stoneflies are living there to determine the best color options for your flies.

Giant Stonefly

Sofa Pillow

page 201

aka Eastern Salmonfly
Pteronarcys biloba, P. dorsata
Western NY: third week of May (sporadic adults may appear much earlier)
Catskills: fourth week of May (sporadic adults may appear much earlier)
Adirondacks: second week of June (sporadic adults may appear much earlier)
Duration: one week
afternoon through evening or early morning
#4–8 (23–40 mm)

Description

Nymph: Black to dark brown tails, body, and legs; black wing cases mottled with brown. *P. biloba* has prominent spines on its abdominal segments while *P. dorsata* does not; prominent tufts of gills line the underside of the thorax of both species; underside usually

Above left: Giant Stonefly nymph (*P. biloba*); *Above:* Underside of Giant Stonefly nymph, showing gill structure and belly color; *Left:* Female Giant Stonefly adult (notice her upturned abdomen for carrying the egg sac)

tannish-orange, much lighter than its back. **Adult:** Black tails and legs; orange body mottled with black; light dun wings with black venations. Resembles Golden Stonefly.

Emergence

Giant Stoneflies emerge when oxeye daisies and blue violets are blooming and spring *Isonychia* are hatching, though you may see a few scouts as much as a month earlier. They are sensitive to light and often appear near dusk or at night, but they can also emerge during the day, particularly in forested areas.

Behavior and Tactics

Dead-drift black Stonefly Nymphs throughout the year, whenever the water is high or off-color in rivers and streams that contain *Pteronarcys* hatches. Though the emerging nymphs are large, Giant Stoneflies have a three-year life cycle, so nymphs are available to trout in various sizes throughout the year. Slowly strip nymph patterns toward the banks during the hatch's peak to imitate nymphs that are moving toward riverside rocks to emerge.

Giant Stonefly adults aren't always available to trout because they often emerge on exposed rocks, but they emerge in-stream if necessary. Giant Stoneflies are clumsy fliers. Fishing on windy days, or in river sections with overhanging vegetation, will increase your chances of finding trout eating Giant Stonefly adults that are blown, or fall, into the river. Blind casting large Sofa Pillows or Stimulators can be effective at this time. Use heavy tippet and prepare yourself for violent strikes if you skitter your fly across a riffle.

Mating takes place on dry ground, usually on rocks or in streamside vegetation, and females sometimes lay their eggs by dipping their abdomens into the water while standing on the shore. But some Giant Stones lay their eggs by dropping to the water to release them, and blind casting drys is tremendous at these times.

Notes

The largest stonefly species in North America, *Pteronarcys dorsata* seldom emerges in large numbers and is therefore often overlooked by anglers. Because of their great size, many anglers associate them with only large rivers and creeks, but some small streams also have fishable hatches.

Though *P. dorsata* is the most often discussed Giant Stonefly hatch (probably because of the adults' impressive size), populations of *P. biloba* may be more common throughout New York.

Roachlike Stonefly

aka Roachfly
Peltoperla and *Tallaperla* species
Western NY: first week of June
Catskills: second week of June
Adirondacks: third week of June
Duration: three weeks
midday through afternoon
#14–16 (9–12 mm)

Yellow Elk Hair Caddis

page 201

Description

Nymph: Reddish-brown body, tails, and legs. White stringlike gills. Three oversized plates on the back (two for wing cases) that give the nymph the appearance of a terrestrial roach **Adult:** Dirty-yellow tails, body, and legs; light gray wings with prominent black venations. Adults look like small Golden Stonefly adults.

Emergence

Roachlike Stoneflies live in cold, clean, swift-flowing creeks and rivers throughout New York. Though these insects prefer to hatch in low light, sporadic daytime emergences are common because the streams in which they live are generally shaded by dense forest canopy. They seem to prefer stream habitats similar to those of Yellow and Green Sallies. June emergences are most common in New York State.

Behavior and Tactics

Roachlike Stonefly nymphs are shredders that eat decomposing leaves, and they are most often found in leaf packs trapped between stones in riffles. Because they are often buried in the leaves, they may not always be available to trout, though Leiser and Boyle in their book, *Stoneflies for the Angler,* mention trout nosing along leaf packs looking for the nymphs. During high-water events, the nymphs become more readily available to trout as they are flushed out of their leaf packs. A reddish-brown Hare's Ear, tied very wide at the thorax, can produce fish at this time. Because the adults look just like a small Golden Stonefly, you don't need to tie an elaborate dry-fly imitation. The same yellow Stimulators that work for imitating Golden Stones work just as well for Roachlike Stonefly adults when tied in smaller sizes. A large yellow-bodied Elk Hair Caddis is also an effective imitation.

Notes

Though their name sounds menacing, Roachlike Stoneflies do not eat other insects. They would rather graze on decomposing leaves.

Roachlike Stoneflies can be extremely abundant in streams. It's common to find hundreds of them living among the leaf packs in a single riffle.

You can tell the *Peltoperla* genus from the *Tallaperla* genus by looking for paired light and dark spots on their oversized wing cases. *Peltoperla* has the spots and *Tallaperla* does not.

A bright, yellow-bodied Elk Hair Caddis is a good imitation for small stonefly adults like the Roachlike Stonefly. Tie them on 2XL shank dry-fly hooks or curved fine-wire stonefly hooks.

Yellow and Lime Sallies

Yellow Humpy

page 201

aka Little Yellow Stonefly and Little Green Stonefly
Isoperla and *Alloperla* species
Western NY: first week of June
Catskills: second week of June
Adirondacks: third week of June
Duration: two months
afternoon and evening
#14–16 (8–10 mm)

Description

Nymph: Yellow or olive tails and legs, dirty-yellow or olive body, and yellow or olive wing cases mottled with brown. **Adult:** Bright yellow or lime tails, body, and legs; pale yellow or pale green wings. Yellow and Lime Sallies appear almost identical to each other except for their color and live in the same habitat.

Emergence

These two brightly colored stonefly species live in cold, clean, swift-flowing creeks and rivers throughout the state. They both usually hatch anytime from June to mid-August, beginning their emergence

Yellow Sally nymph

when elderberry and chicory are opening and *Ephemera varia* is hatching. Though these insects prefer to hatch in low light, sporadic daytime emergences are common because the streams in which they live are generally shaded by dense forest canopy.

Behavior and Tactics

Yellow and Lime Sallies have a one-year life cycle. Their nymphs usually migrate toward shore and crawl to dry, in-stream rocks to emerge. Yellow and Lime Sallies seldom appear in large numbers, and they are often ignored by trout in waters with abundant aquatic insect populations. The best approach for fishing this hatch is to blind-cast sparsely tied Stimulators or Elk Hair Caddis in mountain brook and brown trout streams. These streams are generally less fertile, so the stoneflies are an important food source for trout.

Notes

Lime Sallies are generally less common and less prolific than Yellow Sallies.

The nymph in the photograph has oddly shortened tails. They appear to be broken. The tails normally taper down to a fine point. I may have inadvertently broken the tails while collecting the specimen.

Though the Yellow Humpy is an upright-winged pattern and is often used as a mayfly attractor pattern, it works well for catching brook trout when Yellow Sallies are hatching. The fish seem to key on the bright yellow body and ignore the upright wings. But the wings make the fly more visible to the angler when fishing the high turbulent streams where brookies usually live. You can also tie the fly with a lime body to imitate Lime Sallies.

6 Other Important Aquatic Insects and Crustaceans

Crane Fly

Diptera, Tipulidae
#6–20 (5–50 mm)

Parachute Dry Fly and Walt's Worm

(See appendix for recipes.)

Spider (Adult, tied by the Dette shop)

page 201

Description

Larva: Cigar-shaped worms in various shades of cream, tan, brown, and olive. **Adult:** Six long legs and two small wings that they generally keep folded over their backs. They are poor flyers and are often easy prey for birds and animals, including trout. New York crane fly adults are 5 to 50 millimeters long, and can live up to 15 days after hatching from larvae. The most prominent body colors for the adults are yellow, olive, and orange. Resemble midge and mosquitoes, though crane flies do not bite humans.

Emergence

Crane flies may emerge anytime during the day from April to November, but they usually hatch in the afternoon during colder weather. As the weather and water warm, expect to find them in the morning and evening. Best hatches appear when dame's rocket first blooms and often coincide with Big Sulphur hatches.

Behavior and Tactics

In rising or high water, dead-drift Walt's Worms to imitate dislodged larvae. While crane fly larvae are a trout favorite in most

Crane fly larva

Giant Crane fly larva

rivers and streams, the adults are usually important only in streams with less diversity of aquatic insect life. In healthy rivers, where aquatic insect life is diverse and abundant, trout often ignore adult crane flies, especially if the crane flies are hatching simultaneously with other aquatic insects, since the crane fly adults offer little protein in their giant legs, prominent wings, and proportionately small bodies. Adults are poor flyers and often fall onto the surface of the water. Exact imitations are a fly-tying challenge, but fortunately are not always necessary—any skittered fly such as a Spider or Variant works well. Oversized hackle parachute dry flies can take trout eating crane fly adults in slow pools.

Notes

Crane flies, a vast group of mosquitolike insects that live in most New York trout streams, are the largest family of true flies. Approximately 1,600 known crane fly species have been identified in North America. Of these, 107 are known to live in aquatic habitats as larvae, making them potentially important to fly fishermen. They are most often found in clean rivers and streams lined with trees. Some species also live on or near stream banks and may be washed into the water during high flows associated with runoff and heavy rains. Egg-laying crane flies can appear in massive numbers at dusk. Many anglers miss these insects on the water, believing that the trout are feeding on spent mayfly spinners.

Two invasive European crane fly species (*Tipula oleracea* and *Tipula paludosa*) are wreaking havoc in New York by destroying turf grass in yards and golf courses. Only four species of crane flies in North America (two indigenous and two invasive) are known to cause crop or grass damage.

Midge

Adult midge

Diptera, Chironomidae
Year-round
#16–32 (2–15 mm)

Wet and dry midge patterns

Red Zebra Midge, Rainbow Warrior, Black Zebra Midge, Wes' Angel Case Midge, Hi-Vis Midge Adult (See appendix for recipes.)

Description

Larva: Wormlike bodies are slightly tapered toward tails, not cigar-shaped like crane flies. Found in a wide range of colors, with black, olive, cream, yellow, gray, and red the most common. **Pupa/Adult:** Pupae are shorter, chunkier versions of the larvae, with prominent

Red Midge larva (aka bloodworm)

Black Midge adult

thoraxes that house the adult's folded wings. Adults have six long legs and two small light-colored wings that they generally keep folded over their backs. Male midge adults often have prominent, bushy antennae. Resemble crane flies and mosquitoes.

Emergence

Midges hatch all year, though warm afternoons when the weather is typically cold in the winter, and few other insects are present, can be an excellent time to find fish feeding on these diminutive insects.

Behavior and Tactics

Some larvae live in small cocoons, others are free-swimming, and some live on the stream bottom just like mayfly nymphs. Pupae live in the same habitats as the larvae. To match dislodged larvae or emerging pupae, dead-drift a Zebra Midge under an indicator or as a dropper on the bend of a dry fly. To imitate pupae emerging just under the surface, dead-drift small emerger patterns that hang just under the film. Sight-fish to rising trout with a single adult midge or a larger dry fly that imitates a cluster of adults, which is easier for you to see and provides the illusion of a larger mouthful for trout.

Notes

Midges have a complete life cycle, just like caddisflies. They undergo four distinct phases: egg, larva, pupa, and adult. They live in nearly every type of water, from lakes and ponds to rivers and creeks and even outflows from sewage treatment plants. Though small, midges provide a dependable year-round food source for trout in many streams. Collect specimens to determine which colors are most prominent in the streams you are fishing. Like mayflies, early- and late-season midges tend to be darker than summer midges.

Midge larva twist and turn in the water, often positioning their bodies in a U-shape. The curved hook of a Zebra Midge fly pattern imitates this behavior very well.

Scud

Tan scud

aka Freshwater Shrimp
Amphipoda
#12-18 scud hooks (5–20 mm)

William's Cress Bug and Tan and Olive Scud

page 203

Scuds, also called freshwater shrimp, have olive, tan, or gray bodies and legs. Capture specimens to find the most prevalent color. Scuds turn an orangish color when they die, and many anglers catch trout with orange scud patterns. Scuds are good swimmers and prefer cold, alkaline waters. To imitate them, dead-drift #12–18 scud patterns through riffles or near pockets of aquatic vegetation or strip them with quick, short strokes to imitate swimming scuds.

SCUDS AND SOW BUGS

Scuds and sow bugs (which are crustaceans) can be important trout foods, especially in tailwaters and spring-fed streams. They spend their entire lives underwater, living beneath rocks or in aquatic vegetation.

Sow Bug

Sow bug

aka Pillbug, Cress Bug
Isopoda
#12-20 scud hooks

Sow bugs live in many of New York's streams. Unlike scuds, which are good swimmers, sow bugs generally crawl along aquatic vegetation—their preferred habitat. They have gray bodies and legs. To imitate them, dead-drift William's Cress Bugs or Walt's Worms through riffles and near pockets of aquatic vegetation.

Sow bugs often live in streams that receive inflows of organic material from farming runoff, and they are sometimes an indicator of water pollution. Sow bugs are generally less prolific than scuds, but trout feed on them where they are found.

Aquatic Earthworms

Aquatic earthworm

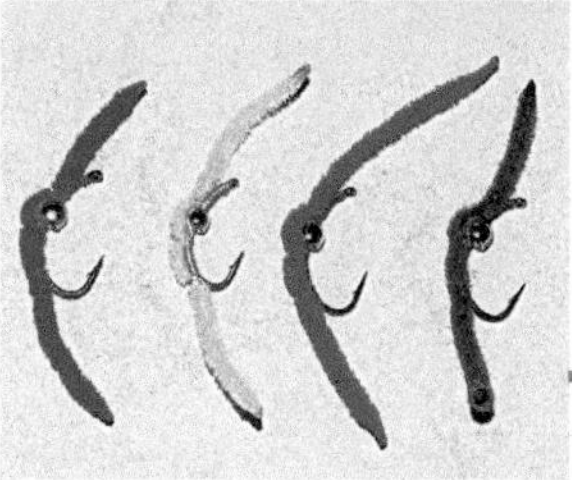

Annelida Oligochaeta
1–30 mm

San Juan Worm

page 203

Aquatic earthworms look very similar to the terrestrial earthworms living in your yard. Though there may be more than 200 species of aquatic earthworms in North America, few have been well studied and little is known about most of them. However, it is known that aquatic earthworms can live in streams with very low oxygen levels, a sign of water pollution. I have found them in nearly every stream and river that I've sampled—including some of New York's best, cleanest waters. They are a very important food source for trout.

Many anglers believe that fishing a San Juan Worm after a rain storm, when the water becomes clouded by runoff, is so productive because terrestrial earthworms are washed into the creek from the banks. That may happen occasionally, but many aquatic earthworms become dislodged in high-water events, becoming easy targets for trout, and that is more likely the true key to the San Juan Worm's effectiveness.

Dobsonfly, Fishfly, and Alderfly

Dobsonfly (hellgrammite adult)

Order Megaloptera
Families Corydalidae and Sialidae

Hellgrammite

page 203

Dobsonflies (the adult form of the hellgrammite) and Fishflies (Corydalidae family) are large insects, from 25 to 90 mm in their larval stage. They generally live in well-oxygenated riffles in the lower reaches of trout streams, where the water tends to be warmer.

Hellgrammite

Fishfly larva

Alderflies (Sialidae family) are smaller, from 10 to 25 mm as larvae, and they are most commonly found in slow-moving water. All of these flies have complete life cycles (including a pupal stage), but most of their cycles are completed out of the water. Because of this trait, pupae and adults are generally unavailable to trout. But their larvae live underwater from one to three years, and trout eat them. The effectiveness of black Wooly Buggers may be due to the pattern's close resemblance to Megaloptera larvae.

Trout seldom eat Dobsonflies, but they are slightly more likely to eat Fishfly and Alderfly adults. These two species tend to be found more often on the water's surface than Dobsonflies, though Alderfly adults sink fairly quickly. It may not work every season, but blind casting Meck's Fishfly dry can produce trout when you see Fishflies in the streamside vegetation. It's also a good fly to use as

Adult Fishfly

Charlie's Fish Fly Adult (Charlie Meck)

page 203

Alderfly larva

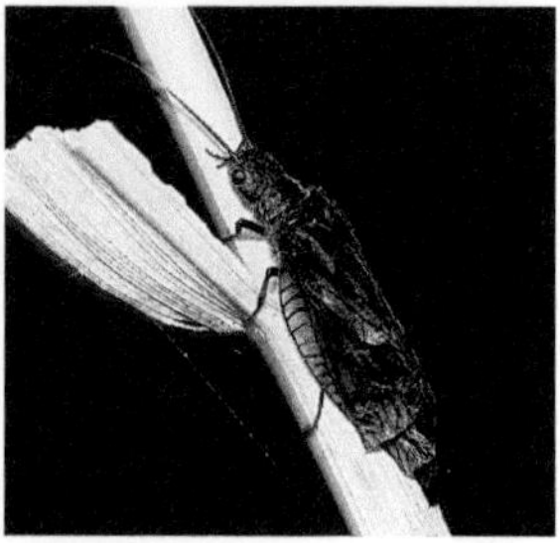

Alderfly adult

Black Trude (Dette Shop)

page 204

Ed Bendel Wet Fly (Ed Bendel)

page 204

the dry in a tandem dry-and-dropper nymph rig when Fishflies are found in streamside vegetation.

Notice the gills along the abdominal appendages of the hellgrammite and compare it to the Fishfly larva. Only the hellgrammite has gills beneath its appendages. The prominent gills are the easiest way to differentiate the two insects in their larval forms.

The Ed Bendel Wet Fly pictured here is a great pattern to fish when Alderflies are hatching. Mr. Bendel was a well-known Adirondack fly tier and fly shop owner in Northville, New York. He was regionally famous, receiving little national recognition until *Fly Tyer* magazine featured him several years ago. Sadly, Mr. Bendel has now passed away. My wife and I stopped at his shop years ago, where I had the opportunity to meet Mr. Bendel and purchase the wet fly photographed here. He was a very kind man and a wonderful fly tier.

Damselflies and Dragonflies

Order Odonata

Blue Damsel

page 204

Damselflies and dragonflies have incomplete life cycles. Their nymphs, which range from 10 to 60 mm, begin hatching in late spring. They are commonly found in lakes and ponds, but they also live in slow-moving sections of some New York trout streams and rivers. The Upper Delaware River has a particularly large population of damselflies and dragonflies. It's common to see the adults dive-bombing the river in late May and early June, eating Sulphurs and Green Drake duns. These adults are generally unimportant to New York fly anglers because trout do not eat them often, but there are rare times when enough of them reach the water's surface, usually

Above left: Dragonfly larva; *Above:* Damselfly larva; *Left:* Damselfly adult

during heavy emergences, that trout do eat them. Trout regularly eat the nymphs when they begin migrating toward dry streamside vegetation, logs, or rocks to emerge.

While collecting *Litobrancha recurvata* duns and spinners in May of 2012, I spent an afternoon watching and photographing a dragonfly nymph as it crawled from a small brook onto a blade of grass and transformed to an adult. It took about 1 hour and 45 minutes from the time the adult began emerging until it flew away from its empty shuck.

7

Terrestrials

The term "terrestrial" is used to encompass all land-dwelling insect species. These insects do not hatch in trout streams, so emerger fly patterns are unnecessary. Because terrestrials do not live in water, they must either fly, fall, hop, or be blown into the water for fish to have the opportunity to eat them. Terrestrials are important in all of New York's trout streams, but they are particularly vital to trout living in streams with fewer aquatic insects.

The most prominent color of each terrestrial species listed in this section can vary greatly from region to region and stream to stream. Match specimens in your area.

Flying ants

Ants and Flying Ants

Hymenoptera Formicidae
Flying ants from August 20 to October 30
afternoon
#12–24 (2–12 mm)

CDC Ant

page 204

Description

Ants can have black, red, or cinnamon bodies, or any combination of these colors. Wings on flying ants are clear or light dun colored. Flying ants resemble termites, but the ants have three distinct body parts whereas termites have only two.

Flying ant

Emergence

Land-bound ants are available to trout from late spring through the summer, until the first or second hard frost. Flying ants can usually first be found on New York trout streams within three days of August 25. They can appear any time after their initial flight, but they seem to prefer dry, hot, breezy weather. Flying ants appear just as the goldenrod is blooming and often at the same time as the White Fly hatch.

Behavior and Tactics

Ant colonies are made up of queens, workers, and male and female winged ants that have the sole purpose of reproducing to populate other colonies. Worker ants are sterile female ants who engage in the day-to-day activities (just about all of the activities) of tending to the queen, building and maintaining the nest, gathering the food, and protecting the colony. The females going about their work are the ants that we see most of the time.

At certain times of the year, the queen produces male ants and female ants that have wings. Male ants, also called drones, do nothing but eat and mate. When it is time to mate, the winged males and females pour out of the ground or through cracks in your sidewalk and take to the air. As the ants take to the sky, they meet and mate in swarms. Shortly after mating, the male ants die. When these ants swarm over water, the dying ants falling to water's surface can bring up all the trout in a given stretch. The females then land, drop their wings, and start new colonies.

Fish ants near overhanging vegetation where ants can fall or get swept into the water by the wind. Ants, both flying and nonflying, should be fished with a dead drift, near the shore, when prospecting for trout.

Grasshoppers

Grasshopper

Orthoptera
#6–14, 2XL (15–40 mm)

Dave's Hopper

page 204

Description

Grasshoppers come in a wide range of colors (most commonly green, yellow, tan, and gray) and sizes and have long, thick bodies, two antennae, and powerful, prominent rear legs. Resembles katydids, which are large and bright green.

Emergence

Grasshoppers are available to trout from late spring throughout the summer, until the first or second hard frost. They are found all over New York, but are particularly plentiful along streams adjacent to fields or grassy areas within forests. You should usually start fishing hoppers when milkweed is blooming or as soon as you see them hopping away from you as you walk through a meadow.

Behavior and Tactics

In the early morning, hoppers are less active than they are at midday, so it's usually more productive to give the sun some time to warm things before you begin casting hopper imitations. Fish Para-Hopper, Dave's Hopper, or Joe's Hopper dry-fly patterns close to stream banks, where a real hopper could inadvertently jump or fly into the water. Cast the flies with enough force to create a small, rippling disturbance when they hit the water. It's also effective to fish hoppers as the lead fly in tandem with small to medium-sized nymphs, using the grasshopper as a strike indicator.

Notes

Immature grasshoppers of larger species as well as small species of grasshoppers are also available to trout. Small grasshopper patterns (#12-14, also imitated by large caddis patterns or small Stimulators) can work very well. Many stonefly patterns do double duty as grasshopper patterns, and the fish may take hoppers for stoneflies, or stoneflies for hoppers. Some grasshoppers can fly great distances, though they usually keep their wings folded over their backs at rest.

Crickets

Orthoptera Gryllidae
#8–12, 2XL (12–25 mm)

Dave's Cricket

page 205

Description

Crickets have long, thick bodies, powerful rear legs, and two prominent antennae. Field crickets are larger than crickets common in the woods. Crickets are usually black or dark brown.

Emergence

Crickets are available to trout from late spring through the summer, until the first or second hard frost. They are found throughout New York, but are particularly plentiful along streams adjacent to forests. Start carrying cricket fly patterns when you hear the male cricket's first distinct chirp in the evening.

Behavior and Tactics

Crickets are more active in low light than grasshoppers. For anglers, crickets are the woodland version of the grasshopper (though crickets and grasshoppers can often be found occupying the same habitat). Fish them like you do hoppers, casting them at likely spots with a pronounced splat. A Dave's Cricket or Letort Cricket imitation works very well for prospecting with a dry fly.

Notes

Many anglers report better success with cricket imitations than grasshoppers on New York streams. The Letort Cricket, designed by Ed Shenk, is a superb cricket pattern that can also be adapted to imitate grasshoppers.

Only male crickets can chirp, and they do so to attract a mate. They produce the sound by rubbing their wings together.

Beetles

Japanese beetle

Coleoptera
#12–20 (3–20 mm)

Dette Beetle (the Dette shop)

page 205

Description

Beetles generally have rounded or elongated oval bodies and two pairs of wings that they keep tightly folded on top of their backs when at rest. Many of the most common angler-significant beetles near trout streams are black with hints of green on their undersides. But the Japanese beetle, which can be prolific along some trout waters, has a metallic green head and legs. Its underside is black, and its wings are metallic copper.

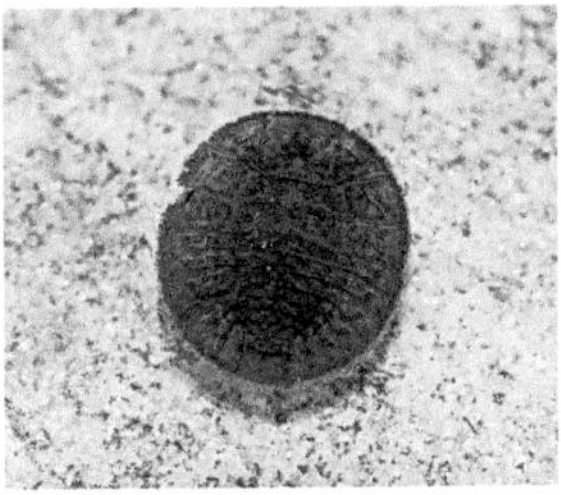

Though its shape makes it difficult to believe, this Water Penny is an aquatic beetle larva.

Beetle come in all shapes, sizes, and colors. This Tiger Beetle, a ferocious predator, is bright emerald-green.

Emergence

Beetles are available to trout from late spring through the summer, until the first or second hard frost. They are plentiful throughout all regions of New York. Japanese beetles are active when spotted Joe Pye weed and bull thistle are blooming and when gypsy moth caterpillars change into adults. Some beetles are land born and must fly or fall onto the water to be accessible to fish. But other species, such as the Water Penny, Riffle Beetle, and others, begin life as aquatic larvae.

Behavior and Tactics

Cast a Hi-Vis Crowe beetle or Dette Beetle close to a stream bank or in-stream structure, such as a downed tree, with enough force to create a gentle plop when it lands on the water's surface. Expect trout to strike the fly as soon as it hits the water. During the summer and early fall, fish a larger beetle pattern (#10-12) as an indicator fly with a small nymph dropped off the bend.

Notes

Beetles are the largest order of insects in the world and the most prolific life form on the planet. In addition to black beetles of all sizes, lightning bugs, Japanese beetles, and ladybugs can all be important trout foods.

Many beetles are aquatic and live much of their lives in the water. Sunken adult and larval beetle patterns can be tied to imitate these species.

Though Japanese Beetles look great in photographs, standard black beetles are much more common along New York trout streams.

While most beetle fly patterns are black, the Dettes tie a version with white deer hair. This fly is much easier for the angler to see on the water. It may not work all the time, but if you're having a hard time seeing your fly, try a white Dette Beetle.

Cicadas

Dog Day Cicada: *Tibicen*
Periodical Cicada: *Magicicada septendecim*
#6–10 (2–5 cm)

Hoover's Periodical Cicada

page 205

Periodical Cicada

Description

Cicadas are large insects that have two transparent wings with prominent veins (green or brown veins for Dog Day and orange for Periodical) that are kept folded over their backs when the insects are at rest. Eyes are olive to green in the Dog Day Cicada and bright reddish orange in the Periodical Cicada. Varying eye colors like gray, blue, or brown are also possible, but rare, for the Periodical Cicada. Bodies and legs are mottled black, brown, olive, and green for the Dog Day and black, brown, and reddish orange for the Periodical.

Emergence

Dog Day Cicadas have a 2- to 5-year lifespan, and they are slightly larger than Periodical Cicadas. They spend most of their lives beneath the ground, feeding on fluid from tree roots, and emerge from mid July through late August by digging a hole to the surface. Once they emerge, they generally live as adults for 3 to 4 weeks before mating, laying eggs, and dying. Only males "sing" to attract a mate, and they do so by vibrating a membrane in their abdomen, not by rubbing their wings together like a cricket. Cicadas like the heat of the day and are most active when it's hottest. After mating, females saw into tree limbs to deposit their eggs. After the nymphs hatch, they crawl down the tree and dig into the soil to find tree roots and begin the cycle anew.

Periodical Cicadas have the same life cycle as Dog Day Cicadas, but it takes them longer to complete it. The Periodical Cicada life span is either 13 or 17 years. There are 6 Periodical Cicada species: three 13-year species and three 17-year species. These 6 species have 17 broods, which are cataloged with Roman numerals from I through XVI. Periodical Cicadas usually appear from late May through the beginning of July and live for about 3 to 4 weeks as adults.

Notes

Cicadas aren't always important for fly fishing. But trout love to eat them if there are enough to reach the water. It is worth fishing Cicada patterns as the drys in dry-and-nymph-dropper rigs throughout the summer. The foam flies are very buoyant and visible, making them excellent strike indicators, and you just might find a trout that wants to eat it. Dry-fly fishing with Hoover's Periodical Cicada pattern can be phenomenal in years with large numbers of Periodical Cicadas emerging. For fishing during Dog Day Cicada years, simply alter Hoover's cicada pattern by changing the colors from orange to green.

There are about 2,500 cicada species around the world. They are often called locusts; however, they are unrelated to the grasshopper-like insect. Cicadas are more closely related to leafhoppers.

Cicadas aren't just food for fish. Birds, bears, squirrels, and just about every other creature with access to them eat cicadas. There are even people that eat the soft, white Periodical Cicadas when they first appear from the ground. This dish was a favorite of Native Americans.

Periodical Cicadas are the oldest living insect in North America.

You can see the empty nymphal shuck behind the adult in the Periodical Cicada photo.

Moths and Caterpillars

Lepidoptera Geometridae (and others)
#8–16, 2XL (10–55 mm)

Muddled Head Moth and Green Weenie

page 205

Description

Caterpillars are thick-bodied, wormlike creatures that are usually covered in fine, stiff hairs. They can be many different colors, depending on their species and diet. Inchworms have smooth, nearly hairless bodies that are often green, brown, or black. Moths (adults) are often gray, brown, or a powdery tan color, but they can have a wide range of colors, depending upon species. They have

Inchworm

Moths can be many different shapes and colors. This Rosy Maple Moth is not necessarily a prime trout food. It is shown here to provide a general body form for many moth species.

broad wings and thick, substantial bodies that are often the same color as the wings. Anglers sometimes mistake moths for caddisflies.

Emergence

Caterpillars and moths, including inchworms, gypsy moths, and forest tent moths, are common throughout New York's forest regions. Their transformation from caterpillar to moth is usually a summer event. Inchworms dangle from tree limbs in mid-May and June. The adult moths are most active early on summer mornings and at night. Gypsy Moth caterpillars are active when wild geranium is blooming, and the adult moths, when the wild sarsaparilla blooms.

Behavior and Tactics

Inchworms can be prolific along trout streams where they dangle precariously from tree limbs by their silk lines. Many inchworms fall into the water, providing trout the opportunity to eat them. Green Weenies and other sunken inchworm patterns fished under a dry-fly indicator work well from late spring through summer. Wooly Worms can be tied in various colors to imitate hairy caterpillars.

Large caddisfly patterns work well for most caterpillar and moth adults. Fish them with a dead drift or an occasional subtle twitch.

The naturals may be unimportant in some regions, even those where moth and caterpillar infestations are heavy.

Though trout eat caterpillars and inchworms every season, moth fishing is much less reliable. The best moth fishing that I've experienced came during a heavy Forest Tent Caterpillar infestation along the Delaware River's Upper East Branch. Icy-cold water releases from the Pepacton Reservoir created dense fog, which seemed to stun the moths, making them fall to the water. The fish gorged on these tan moths for nearly two weeks, instigating some of the best dry-fly fishing of my life.

Appendix: Fly Patterns

CHOOSING THE RIGHT FLY AND TACTICS FOR FISHING IT

Having a solid grasp of the hatches will not help you catch more trout if you don't know how to imitate the real bugs with artificial flies. Hundreds of books have been written about selecting flies and techniques for fishing them. If you are overwhelmed by all the information that is available, my advice is to take bits and pieces from what you read—the parts that make the most sense to you—and work on perfecting those elements. No one has the single right fly or best technique to use every day. That pattern and technique doesn't exist.

Selecting the right fly pattern takes time to master, but here are a few important considerations.

Location. Where in the water column are the naturals floating; where are the fish feeding? If the fish are feeding just under the surface, and showing the bulging rises characteristic of that, then you are better off fishing a low-riding fly or an emerger that hangs beneath the surface rather than a heavily hackled pattern that rides high on the water.

Pressure. For unpressured fish and mountain streams, you can get away with a few different flies that represent the general shapes of mayflies and caddis. A good fastwater fly should float well and be durable and easy to see.

Water Type. In fast water and riffles, a hackle pattern floats higher and the fish are often less selective. In slow water, flush-floating patterns are usually a better choice.

Time of Day. In the bright light of midday or afternoon, you may need to use a sparser, more accurate pattern than you would while fishing a hatch or spinnerfall in low light. On bright days, especially when the water is clear, you might also have to fish longer leaders and more delicate tippets. Fish as short a leader and as strong a tippet as you can get away with.

Weather. If it is windy and duns are being blown across the water's surface, a hackled pattern that can skitter on the water may outperform a flush-floating one. If it is cold, duns generally take longer to take flight, and a flush-floating pattern works well. When the weather is warm, mayflies move more, which can be better imitated with a hackled pattern. During low light conditions, a high-riding pattern or one tied with a highly visible parachute post has a distinct advantage over a drab pattern that rides closer to the surface. In the silver sheen of sun glare, a fly with a black wingpost is easier to see than one with a white post.

Fish-Eye View. When choosing a pattern, consider what the fish sees. When trout are feeding on mayfly duns, they do not care about the color of the back of the bug. They see only the belly, and that is what you want to match. Consider the components of your imitation from below. Is the body the right size? Is the wing profile accurate?

The following chart outlines how I normally approach hatches, pattern selection, and fishing styles during various weather and water conditions. There are no hard-and-fast rules to fly pattern selection or fishing tactics, and sometimes breaking the suggestions in the chart below catches fish. But the information that I've included is a great place to begin your day. Most of the time, this system works for me.

The Dette Fly Shop in Roscoe, New York, is the oldest continuously family-owned fly shop in the United States.

SELECTING A FLY PATTERN

Fly style	When the water is high and/or dirty	When the water is low and/or clear	Fishing when it's cloudy or rainy	Fishing i the brigh sun
Non-bead-head nymph		•		•
Bead-head nymph	•		•	
Nymphs tied with synthetic flash	•		•	
Nymphs tied with mostly natural materials		•		•
Wet fly	•		•	
Parachute dry fly/ Truform Fly, Comparachute		•		•
Comparadun, Sparkle Dun, CDC Comparadun-style dry fly		•	•	•
Catskill-style dry fly and wulffs	•		•	
CDC, snowshoe, or deer hair emerger dry fly		•	•	•
Hackled caddis dry (Elk Hair Caddis, Trude, Dette Caddis)	•		•	
Non-hackled caddis dry (CDC or X Caddis)		•		•

• Designates that this is a good time to fish this style of fly.

ishing at dusk or dark	Insect emerges on stream bottom	Insect emerges near the surface	Fishing riffles	Fishing pools
		•	•	•
•	•		•	
•	•		•	
	•	•	•	•
•	•		•	
		•		•
		•		•
•	•		•	
		•		•
•	•	•	•	
	•	•		•

Mary Dette Clark and Joe Fox own and operate the Dette Fly Shop in Roscoe.

NEW YORK FLY SHOPS AND FLY DESIGNERS

Many of the fly patterns in this book were designed and developed in New York, which historically has been one of the most influential states for American fly pattern development. Dozens of fly patterns in shops around the world were originally created for catching fish in cold Catskill and Adirondack trout streams and even the lesser-known streams of western New York.

American dry flies initially became popular in New York's Catskill Mountains thanks to Theodore Gordon. Though Gordon was born in Pennsylvania (another hotbed of fly pattern design), he spent much of his adult life living in the Catskills, fishing the Neversink, Beaverkill, and Delaware rivers. England's dry-fly fishing authority, Frederick Halford, sent Gordon a package containing early English dry flies in the late 1800s. Gordon used these flies as a template for developing his own drys to imitate American aquatic insects. His writing in *Forest and Stream Magazine* spread his designs and theories and encouraged more Americans to tie dry flies for their home waters.

The Catskill tiers who came after Gordon are part of a golden age of American fly design that continues today. Rube Cross and a host of others refined Gordon's patterns, developing the Catskill-style fly, which was further refined by Harry and Elsie Darbee and Walt and Winnie Dette. In 2013, the Dettes' daughter, Mary Dette Clark, and her grandson, Joe Fox, are still tying and refining Catskill-style flies at the Dette Fly Shop in Roscoe, New York.

New York has contributed much more to fly design than just the Catskill-style dry.

Art Flick gave us the Variant, a quill-body fly designed for fishing the riffles and runs of the Schoharie. Edward Ringwood Hewitt designed the Skater for the Neversink. Fly-fishing legend Lee Wulff, designer of the Wulff dry flies, spent his final years with his wife Joan on the Upper Beaverkill teaching hundreds to fly-fish at the Wulff School. Al Caucci and Bob Nastasi developed the Comparadun, popularizing it through their books *Hatches* and *Hatches II* and in their legendary school on the banks of the Delaware's West Branch just across the river from New York state.

Not all of New York's fly pattern innovation originated in the Catskills.

Fran Betters, Ray Bergman, Ed Bendel, and a host of others developed fly styles and refined fly patterns in the Adirondacks. Fran Betters developed the Ausable Wulff and Bomber, the Usual, and the Haystack, which was the inspiration for Caucci and Nastasi's Comparadun. Mr. Betters's iconic Adirondack Sport Shop—with its red-skirted cowgirl holding a fly rod hooked to her skirt painted on its front—is a near-holy place for anglers traveling to Wilmington, New York, even though Betters has passed on.

New York state's deep fly-fishing traditions continue today with innovative tiers who refine the old standards and break new ground with modern fly patterns. Fly shops have never been more abundant in New York, a mix of the old and the new, from the quaint, historic Dette Shop to major modern operations like West Branch Angler & Resort in Deposit, New York.

The fly patterns featured throughout this book and in the following list, are a mix of the old and the new, a continuation of New York's amazing fly-fishing legacy.

West Branch Angler & Resort as seen through the fog on an early-morning Delaware West Branch float trip.

BASIC MAYFLY NYMPH PATTERNS

Hare's Ear

Hook:	#6-22 standard or 2XL nymph or heavy-wire scud (for curved nymphs)
Thread:	Black 8/0 Uni
Bead (optional):	Gold, black, silver, or copper
Tail:	Brown woodchuck or partridge
Body:	Natural, ginger, brown, or olive hare's ear dubbing
Rib (optional):	Flat gold tinsel
Wing case:	Mottled turkey wing quill

Bead Head Flash Back Pheasant Tail

Hook:	#6-22 standard nymph or heavy-wire scud (for curved nymphs)
Bead (optional):	Gold, black, silver, or copper
Thread:	Dark brown 8/0 Uni

Tail:	Pheasant tail fibers
Abdomen:	Pheasant tail fibers
Rib (optional):	Fine oval copper wire
Thorax:	Peacock herl
Wing case:	Mottled turkey wing quill
Legs (optional):	Pheasant tail fibers

Copper John

Hook:	#6-18 standard nymph
Bead:	Gold
Thread:	Rust 8/0 Uni
Tail:	Black goose biots
Abdomen:	Copper wire
Thorax:	Peacock herl
Wing case:	Wapsi Thin Skin and a strip of Flashabou coated with epoxy
Legs:	Partridge or henback

Beadhead Prince

Hook:	#6-18 standard nymph or heavy-wire scud (for curved nymphs)
Bead (optional):	Gold, black, silver, or copper
Thread:	Black 8/0 Uni
Tail:	Brown goose biots
Rib (optional):	Oval gold tinsel
Body:	Peacock herl
Wing case:	White goose biots
Hackle:	Brown

HATCH-MATCHING MAYFLY NYMPH PATTERNS

Humphrey's Sulphur Nymph

Hook:	#14-18 standard nymph
Thread:	Camel 8/0 Uni
Tail:	Partridge dyed brown
Abdomen:	Pale olive dubbing
Thorax:	Rusty brown dubbing
Rib (optional):	Fine gold wire
Legs:	Partridge dyed brown

William's Dark Mayfly Nymph

Hook:	#12-18 standard nymph
Thread:	Camel 8/0 Uni
Tail:	Mallard flank
Abdomen:	Peacock herl
Thorax:	Rusty brown dubbing
Rib (optional):	Fine gold wire
Wing case:	Black turkey tail
Legs:	Mallard flank

Drake Nymph

Hook:	#4-10, 2 or 3XL nymph
Thread:	Camel 8/0 Uni
Tail:	Brown partridge
Abdomen:	Rusty brown dubbing
Thorax:	Pale yellow dubbing or yarn
Gills:	Ostrich herl
Rib (optional):	7X monofilament
Wing case:	Black turkey tail
Legs:	Brown partridge

Stroup's Slate Drake Nymph

Hook:	#10-14, 2XL nymph
Thread:	Wine 8/0 Uni
Tail:	Mallard flank
Body:	Maroon Ice Dubbing
Gills:	Gray ostrich herl or trimmed saddle feather
Strip:	White 3/0 Uni or 1 strand of pearl Flashabou
Wing case:	Peacock herl
Legs:	Brown hackle

March Brown Nymph

Hook:	#10-14 2XL nymph
Thread:	Camel 8/0 Uni
Tail:	Ginger ostrich herl
Body:	Cream Superfine Dubbing
Gills:	Ginger ostrich herl or trimmed saddle feather

Wing case:	Mottled turkey
Legs:	Furnace hackle

MAYFLY SPINNER PATTERNS

Poly Wing Spinner

Hook:	#8-24 standard or 2XL dry fly
Thread:	8/0 Uni (colored to match the natural's body)
Tail:	Microfibetts (colored to match the natural's tail)
Wing:	Gray polypropylene yarn
Body:	Beaver dubbing (colored to match the natural's body)

Hackle Wing Spinner

Hook:	#8-24, standard or 2XL dry fly
Thread:	8/0 Uni (colored to match the natural's body)
Tail:	Medium dun hackle fibers
Wing:	Badger, cream, or grizzly hackle
Abdomen:	Beaver dubbing (colored to match the natural)
Thorax:	Peacock herl

CDC Wing Spinner

Hook:	#8-24, standard or 2XL dry fly
Thread:	8/0 Uni (colored to match the natural's body)
Tail:	Medium dun hackle fibers or Microfibetts
Wing:	White or medium dun CDC
Abdomen:	Turkey biot or beaver dubbing (colored to match the natural)
Thorax:	Foam (colored to match the body)

MAYFLY DUN AND EMERGER PATTERNS

Weamer's Truform Emergers

Hook:	#8-14 Daiichi 1230
Thread:	8/0 Uni (colored to match the natural's body)
Tail:	Dark brown Darlon
Wing:	CDC (colored to match the natural)
Body:	Beaver dubbing (colored to match the natural)
Trimmed hackle post:	Antron yarn (colored to match the natural's body)
Hackle:	Colored to match the natural's legs

Weamer's Truform Hendrickson Emerger

This style of fly can be used to imitate most mayflies by changing the material colors and hook size.

RS2

Hook:	#18-20 standard-shank dry fly
Thread:	Olive 8/0 Uni
Tail:	Light dun Microfibetts
Wing:	Medium dun CDC
Body:	Olive beaver dubbing
Legs:	Medium dun hackle fibers

Conover

Hook:	#16-18 standard-shank dry fly
Thread:	Black 8/0 Uni
Tail:	Cream hackle fibers
Body:	Blended muskrat and red wool or claret beaver dubbing
Hackle:	Golden badger without black edging

Quill Gordon Emerger

Hook:	#10-12 fine-wire scud
Thread:	Dark brown 8/0 Uni
Tail:	Gray Antron yarn fibers
Wing:	Medium dun snowshoe rabbit
Abdomen:	Stripped peacock quill
Thorax:	Dark brown beaver dubbing

Bucktail Body Hendrickson Emerger

Hook:	#12-14 fine-wire scud
Thread:	Pink 8/0 Uni
Tail:	Dark brown Darlon
Wing:	Cream snowshoe rabbit
Abdomen:	Hendrickson pink beaver dubbing
Thorax:	3 pink, 2 olive, and 2 tan bucktail fibers, twisted with the thread to make a rope

DK Hendo Emerger

Hook:	#12-14 fine-wire scud
Thread:	Red 8/0 Uni
Tail:	Medium dun hackle fibers
Wing:	Light dun polypropylene yarn
Overwing:	Medium dun hackle fibers
Body:	Reddish-brown beaver dubbing

Black Quill BB Parachute

Hook:	#14 2XL dry fly
Thread:	Black 8/0 Uni
Tail:	3 black bucktail fibers
Wing:	Gray Antron yarn
Abdomen:	Black beaver dubbing
Thorax:	5 black bucktail fibers, twisted with the thread to make a rope
Hackle:	Black

TCO CDC Sulphur Emerger

Hook:	#14-16 fine-wire scud
Thread:	Yellow 8/0 Uni (colored to match the natural)
Tail:	Wood-duck fibers
Wing:	Medium dun CDC
Abdomen:	Pheasant tail
Ribbing:	7X monofilament
Thorax:	Sulphur-yellow beaver dubbing

Haystack

Hook:	#10-14, 2XL dry fly
Thread:	Camel 8/0 Uni
Tail:	Rusty brown deer hair
Wing:	Rusty brown deer hair
Abdomen:	Light brown or cream beaver dubbing

Biot Body CDC Comparadun

Hook: #18-20 standard dry fly
Thread: Yellow 8/0 Uni
Tail: Light dun Microfibetts
Wing: Medium dun CDC with a couple strands of wood-duck fibers
Abdomen: Sulphur-yellow turkey quills
Thorax: Sulphur-yellow beaver dubbing

Weamer's Green Drake Comparachute

Hook: #8-12 Daiichi 1230, Weamer's Truform
Thread: 8/0 Uni (colored to match the natural)
Tail: Dark brown Darlon
Wing: Olive deer hair
Abdomen: Pale yellow beaver dubbing
Legs: Grizzly hackle dyed olive

Dette Coffin Fly

Hook: #8-12, 2XL dry fly
Thread: White 8/0 Uni
Tail: 3 peccary fibers, divided
Wing: Teal flank
Rib: White 3/0 thread
Body: Trimmed white saddle hackle
Underbody: White poly yarn
Hackle: Golden badger (black centered and black edged)

Weamer's TF Comparadun

Hook: #8-10 Daiichi 1230, Weamer's Truform Mayfly
Thread: Olive 8/0 Uni
Tail: Dark brown Darlon
Wing: Deer hair
Body: Olive beaver dubbing

Weamer's Burke Emerger

Hook: #10-12 fine-wire scud
Thread: Wine 8/0 Uni
Tail: Dark brown mink guard hairs
Wing: Medium dun CDC
Body: Maroon beaver dubbing

Weamer's Brown Drake Flip Flop Fly

Hook:	#8-10 fine-wire scud
Thread:	Olive 8/0 Uni
Tail:	Moose mane
Wing:	Trimmed brown hackle feather
Body:	Light olive beaver dubbing
Legs:	Brown hackle

Hackle Wing BWO

Hook:	#14-16 standard dry fly
Thread:	Olive 8/0 Uni
Tail:	Light dun Microfibetts
Wing:	Medium dun hackle feather
Body:	Insect-green beaver dubbing
Legs:	Medium dun hackle

Usual

Hook:	#12-14 standard dry fly
Thread:	Gray 8/0 Uni
Tail:	Cream snowshoe rabbit
Wing:	Cream snowshoe rabbit
Body:	Gray beaver dubbing or muskrat

Ausable Wulff

Hook:	#12-14 standard dry fly
Thread:	Flourescent Orange 8/0 Uni
Tail:	Woodchuck
Wing:	White calf body hair
Body:	Orange beaver dubbing
Hackle:	Brown and grizzly mixed

Para-Scud

Hook:	#12-14 fine-wire scud hook
Thread:	Olive 8/0 Uni
Wing:	Light dun Antron yarn
Abdomen:	Light olive beaver dubbing
Thorax:	Orange beaver dubbing
Hackle:	Grizzly dyed olive

Cream Variant

Hook: #14 standard dry fly
Thread: White 8/0 Uni
Tail: Cream hackle fibers
Body: Stripped cream center quill hackle stem
Hackle: Ginger and grizzly mixed, tied 4 sizes larger than standard hackle

Irresistible

Hook: #10-12, 2XL dry fly
Thread: Cream 8/0 Uni
Tail: Bleached elk hair
Wing: Bleached elk hair
Body: Bleached elk hair (spun and trimmed)
Hackle: Light dun

Jim's Paralep Emerger

Hook: #16-20 fine-wire scud
Thread: Iron gray 8/0 Uni
Tail: Dark dun Antron
Wing: White Hi-Vis
Abdomen: Twisted dark dun Antron
Thorax: Dark brown beaver dubbing

Budding Emerger

Hook: #10-12 fine wire
Thread: Yellow 8/0 Uni
Tail: Amber Antron
Wing: White foam
Body: Golden yellow beaver dubbing
Hackle: Golden yellow

Trico CDC Comparadun

Hook: #22-26 standard dry fly
Thread: Black 8/0 Uni
Tail: Light dun Microfibetts
Wing: White CDC
Abdomen: Black turkey biot
Thorax: Black beaver dubbing

Trico Antron Wing Female Spinner

Hook:	#22-26 standard dry fly
Thread:	Black 8/0 Uni
Tail:	Light dun Microfibetts
Wing:	White Antron or poly yarn
Abdomen:	White beaver dubbing
Thorax:	Black beaver dubbing

Sunken Trico

Hook:	#22-26 standard dry fly
Thread:	Black 8/0 Uni
Bead:	Black glass or brass bead
Tail:	Light dun Microfibetts
Wing:	White Antron
Body:	Black Superfine Dubbing

Snowshoe Emerger

Hook:	#8 fine-wire scud
Thread:	Yellow 8/0 Uni
Tail:	Amber Z-Lon
Wing:	Cream snowshoe rabbit
Overwing:	Mallard flank dyed yellow
Body:	Yellow beaver dubbing

Yellow Spun Dun

Hook:	#16-18 standard dry fly
Thread:	Olive 8/0 Uni
Tail:	Light dun Microfibetts
Wing:	Deer hair, spun and clipped to look like a Comparadun
Body:	Dirty yellow synthetic dubbing

Hi-Vis White Wulff

Hook:	#12-14 standard dry fly
Thread:	White 8/0 Uni
Tail:	White calf-tail fibers
Wing:	Calf tail dyed chartreuse
Body:	Cream Superfine Dubbing
Hackle:	Cream Badger

Catskill Style

Hook:	#6-8 2XL dry fly
Thread:	Yellow 8/0 Uni
Tail:	Moose mane
Wing:	Mallard flank dyed brown
Body:	Yellow-tan beaver dubbing
Hackle:	Brown

Gray Hackle-Stacker

Hook:	#6-8, 2XL dry fly
Thread:	Gray 8/0 Uni
Tail:	2 black stripped hackle stems
Wing:	Mixed grizzly and brown hackle pulled vertical with a gallows tool
Body:	Gray beaver dubbing

GENERAL CADDIS LARVA FLY PATTERNS

Standard Caddis Larva and Beadhead Caddis Larva

Hook:	#10-22 heavy-wire scud
Bead (optional):	Black, copper, gold, or silver
Thread:	8/0 Uni, to match the natural's body
Thorax:	Rabbit dubbing to match the natural (usually green, olive, or tan)
Rib (optional):	Fine oval gold or copper wire
Head:	Dark gray, black, or brown dubbing (to match the natural)

Walt's Worm and Beadhead Walt's Worm

Hook:	#12-22 standard nymph
Thread:	Tan 8/0 Uni
Body:	Hareline Hare's Ear Plus dubbing or any natural hare's ear dubbing mixed with clear or light dun Antron

Anderson's Bird of Prey Caddis

Hook:	#12-20 heavy-wire scud
Bead:	Gold
Thread:	8/0 Uni, colored to match the natural
Tail:	Natural Hungarian partridge
Rib:	Pearl Flashabou accent
Body:	Rabbit dubbing to match the natural (usually green, olive, or tan)

Hackle:	Natural Hungarian partridge
Thorax:	Peacock herl

Scudaddis

Hook:	#12-20 standard nymph
Bead (optional):	Black, copper, gold, or silver
Thread:	8/0 Uni, colored to match the natural
Tail:	Ginger, tan, or olive Antron
Rib:	Fine wire (color to match the body)
Abdomen:	Ice Dub to match the natural (usually green, olive, or tan)
Thorax:	Peacock Ice Dub
Back:	Clear Scud Back

GENERAL WET-FLY PATTERNS

Starling and Herl (or Partridge and Herl)

Hook:	#10-20 standard wet fly
Thread:	Black 8/0 Uni
Rib (optional):	Fine black wire
Body:	Peacock herl
Hackle:	Starling or partridge

Quill Gordon-Style Wet

Hook:	#10-20 standard wet fly
Thread:	8/0 Uni, colored to match the natural
Tail:	Medium dun hackle fibers
Rib (optional):	Black, gold, copper, or silver fine wire
Body:	Stripped peacock quill or hackle center-feather quills, colored to match the natural
Wing:	Wood duck or mallard flank
Throat:	Medium dun hackle fibers

Partridge and Green

Hook:	#10-20 standard wet fly
Thread:	8/0 Uni, colored to match the natural
Abdomen:	Bright green, yellow, or orange floss or dubbing
Thorax:	Tan or brown dubbing
Hackle:	Natural Hungarian partridge

Standard Wet

Hook:	#10-20 standard wet fly
Thread:	8/0 Uni (color to match the natural)
Tail:	Brown hackle fibers
Rib:	Gold wire
Body:	Dubbing (color to match the natural)
Hackle:	Brown hen hackle
Wing:	Gray duck quill

ADULT CADDISFLY PATTERNS

Arty's Black Caddis

Hook:	#18 standard dry fly
Thread:	Black 8/0 Uni
Wing:	Mottled gray Web Wing
Body:	Black beaver dubbing
Hackle:	Black

CDC Caddis

Hook:	#18 fine-wire scud
Thread:	Black 8/0 Uni
Wing:	Medium dun CDC
Body:	Black beaver dubbing

Peacock Caddis

Hook:	#12-14 standard dry fly
Thread:	Black 8/0 Uni
Wing:	Deer hair
Body:	Peacock herl
Hackle:	Brown and grizzly

MT's Apple Caddis

Hook:	#14-16 standard dry fly
Thread:	Tan 8/0 Uni
Wing:	Ginger CDC
Abdomen:	Apple-green beaver dubbing
Thorax:	Tan beaver dubbing

Bivisible

Hook:	#14-18 standard dry fly
Thread:	White 8/0 Uni
Tail:	Brown hackle fibers
Abdomen:	Brown hackle
Thorax:	White hackle

Olive Chuck Caddis

Hook:	#16-20 standard dry fly
Thread:	Olive 8/0 Uni
Wing:	Woodchuck guard hairs from back of woodchuck
Body:	Olive beaver dubbing
Hackle:	Grizzly and brown mixed

Spun Head Deer Hair Caddis

Hook:	#12-14 standard dry fly
Thread:	Black 8/0 Uni
Body:	Dark gray beaver dubbing
Wing:	Dark dun deer hair
Head:	Spun and clipped dark dun deer hair

Delaware Adams

Hook:	#12-18 standard dry fly
Thread:	Black 8/0 Uni (the Dette family uses white thread)
Wing:	Rounded grizzly hackle tips
Body:	Medium olive wool or dubbing (bright green may also be used) palmered with grizzly
Tail:	Grizzly hackle fibers
Hackle:	Grizzly and brown mixed

CDC Palmered Elk Hair Caddis

Hook:	#10 standard dry fly
Thread:	Orange 8/0 Uni
Wing:	Orange deer hair
Body:	Tannish-orange beaver dubbing
Hackle:	Ginger CDC palmered over body

STANDARD STONEFLY NYMPH PATTERNS

Simple Stonefly Nymph

Hook:	#4-10 curved stonefly nymph
Bead:	Black or gold (color to match the natural)
Thread:	Black or yellow 8/0 Uni (color to match the natural)
Tail:	Black or yellow goose biots (color to match the natural)
Body:	Black or yellow beaver dubbing (color to match the natural)
Wing case:	Dark brown Darlon
Hackle:	Black or yellow hen hackle (color to match the natural)

Kyle's Golden Stone

Hook:	#8-12, 2 or 3XL nymph
Thread:	Tan 6/0 Uni
Tail:	Brown goose biot
Overback:	Mottled Oak Thin Skin
Rib:	Fine gold wire
Body:	Camel Frog Hair dubbing
Wing case:	Black Thin Skin
Thorax:	Yellow Wings-N-Flash dubbing
Back legs:	Pumpkin Fleck Sili Legs
Front legs:	Tan mallard flank

Rob Lewis Realistic Stone (Black)

Hook:	#4-10 curved stonefly nymph
Thread:	Black 6/0 Uni
Tail:	Hackle stems dyed black
Body:	Black foam
Wing cases:	Black foam
Legs:	Black turkey biots
Antennae:	Hackle stems dyed black
Head:	Black foam
Eyes:	Heavy nylon painted black at the ends

Brian Wolfkiel's Woven Golden Stone

Hook:	#6-10 curved stonefly nymph
Thread:	Yellow 6/0 Uni
Tail:	Brown goose biots
Wing cases:	Mottled turkey
Abdomen:	Brown and yellow embroidery floss
Thorax:	Yellow rabbit and gold Antron dubbing (mixed)
Legs:	Brown hackle

Dette Stonefly Nymph

Hook:	#18-22 standard nymph
Thread:	Black 8/0 Uni
Tail:	Black goose biots
Wing case:	Dark gray mallard quill
Body:	Black beaver dubbing
Legs:	Dark dun hackle fibers

ADULT STONEFLY PATTERNS

Black Stimulator

Hook:	#12-18 2XL dry fly
Thread:	Bright red 8/0 Uni (Black may also be used)
Wing:	Elk hair
Overwing:	White deer hair
Tail:	Black hackle fibers with black Krystal Flash on top
Body:	Black beaver dubbing
Palmered hackle:	Black
Rib:	Copper wire
Hackle:	Grizzly

Flat-Water Stonefly

Hook:	#14-16 standard dry fly
Thread:	Brown 8/0 Uni
Wing:	Gray poly yarn
Body:	Brown beaver dubbing
Hackle:	Brown (clipped flat on top)

Jake Villwock's Para-Stone

Hook:	#6-10 curved stonefly dry
Thread:	Brown 3/0 Uni
Tail:	Tan goose biots
Egg sac:	Brown Thin Fly Foam
Rib:	Brown 3/0 Uni
Underbody:	Golden Stone Life Cycle Dubbing
Back:	Yellow Thin Fly Foam
First wing:	Mottled tan Web Wing
Second wing:	Natural deer body hair
Legs:	Yellow/black-barred rubber legs
Parachute post:	Yellow Hi-Vis
Hackle:	Grizzly

Stimulator

Hook:	#6-12 curved stonefly dry
Thread:	Orange, yellow, or green 8/0 Uni
Tail:	Elk hair
Wing:	Elk hair
Abdomen hackle:	Brown
Thorax hackle:	Grizzly
Rib:	Copper wire
Abdomen:	Bright yellow beaver dubbing or floss
Thorax:	Bright orange yarn, beaver dubbing, or floss

Jake Villwock's Golden Foam Stone

Hook:	#6-10 curved stonefly dry
Thread:	Brown 3/0 Uni
Tail:	Tan goose biots
Rib:	Brown 3/0 Uni
Underbody:	Golden Stonefly Life Cycle Dubbing
Back:	Yellow Thin Fly Foam
First wing:	Mottled tan Web Wing
Second wing:	Natural deer body hair
Legs:	Yellow/black-barred rubber legs
Head:	Brown Thin Fly Foam

Sofa Pillow

Hook:	#4-8 2XL dry fly
Thread:	Orange 8/0 Uni
Tail:	Elk hair
Body:	Orange floss or beaver dubbing
Palmered Hackle:	Brown
Rib (optional):	Oval copper wire (fine)
Wing:	Elk hair
Hackle:	Brown (oversized)

Yellow Elk Hair Caddis

Hook:	#14-16 2XL dry fly
Thread:	Yellow
Wing:	Natural deer or elk hair
Body:	Yellow beaver dubbing
Hackle:	Bright yellow
Rib:	Fine gold wire

Yellow Humpy

Hook:	#14-16 standard or 2XL dry fly
Thread:	Yellow or green 8/0 Uni
Wing:	Elk hair
Body:	Bright yellow or lime green synthetic dubbing
Back:	Elk hair
Hackle:	Brown/grizzly mix or hackle dyed yellow or bright green
Tail:	Elk hair tips

Spider

Hook:	#8-12 2XS dry fly
Thread:	8/0 Uni (color to match hackle)
Hackle:	Color to match naturals (4 times normal size)

GENERAL MIDGE PATTERNS

Red Zebra Midge

Hook:	#16-22, 2X-heavy scud
Bead:	Silver
Thread:	Red 8/0 Uni
Body:	Red 8/0 Uni
Rib:	Fine silver wire

Egan's Rainbow Warrior

Hook:	#16-22, 2X-heavy scud
Bead:	Silver
Thread:	Red 8/0 Uni
Abdomen:	Pearl tinsel (medium)
Wing case:	Stub end of pearl tinsel from abdomen
Thorax:	Rainbow Sow-Scud Dubbing

Black Zebra Midge

Hook:	#16-22, 2X-heavy scud
Bead:	Silver
Thread:	Black 8/0 Uni
Body:	Black 8/0 Uni
Rib:	Fine silver wire

Wes' Angel Case Midge

Hook:	#16-22, 2X-heavy scud
Bead:	Clear or pearl glass bead
Thread:	Black 8/0 Uni
Body:	Black 8/0 Uni
Rib:	Fine silver wire
Wing:	Pearl Flashabou

Hi-Vis Midge

Hook:	#16-30 standard dry fly
Thread:	Olive 8/0 Uni
Wing:	Fluorescent pink Antron yarn
Rib:	Fine copper wire
Abdomen:	Olive 8/0 Uni
Thorax:	Olive beaver dubbing

OTHER AQUATIC INSECT FLY PATTERNS

William's Cress Bug

Hook:	#12-16 heavy-wire nymph
Thread:	Tan 8/0 Uni
Body:	Gray Sow-Scud Dubbing
Back:	Clear Scud Back
Rib:	Clear nylon

Tan and Olive Scud

Hook:	#12-16 heavy-wire scud
Thread:	Olive or tan 8/0 Uni
Body:	Olive or tan Sow-Scud Dubbing (picked out for legs)
Back:	Clear Scud Back over pearl Flashabou
Rib:	Copper Ultra Wire
Tails/Antennae:	Olive or brown hackle fibers

San Juan Worm

Hook:	#12-16 heavy-wire scud
Thread:	8/0 Uni (color to match worm)
Bead:	Gold
Body:	Fuchsia, flesh, pink, or worm-brown chenille

Hellgrammite

Hook:	#2-6, 3XL streamer
Tail:	Black marabou
Body:	Black chenille
Hackle:	Dark brown
Legs:	Black round rubber

Charlie's Fish Fly Adult

Hook:	#4-10, 2XL dry fly
Thread:	Iron-gray 8/0 Uni
Wing:	3 separate clumps of deer hair
Body:	Dark gray to black beaver dubbing
Hackle:	Dark dun to black

Ed Bendel Wet Fly

Hook:	#4-12 wet fly
Thread:	Black 8/0 Uni
Tail:	Guinea fowl fibers
Body:	Dark gray to black beaver dubbing
Hackle:	Guinea fowl

Black Trude

Hook:	#10-14 standard-shank dry fly
Thread:	Black 8/0 Uni
Wing:	Black deer hair
Body:	Black beaver dubbing palmered with black hackle
Hackle:	Black

Blue Damsel

Hook:	#10, 2XL dry fly
Tail:	2 grizzly saddle hackles
Body:	Palmered grizzly hackle
Back and head:	Blue foam
Eyes:	Black mono eyes

TERRESTRIAL PATTERNS

CDC Ant

Hook:	#12-24 standard dry fly
Body:	Black beaver dubbing tied in two mounds
Wings:	White CDC feathers
Hackle:	Dark brown or black

Dave's Hopper

Hook:	#6-14, 2XL dry fly
Tail:	Red hackle fibers beneath a loop of yellow yarn
Body:	Yellow yarn palmered with clipped brown hackle
Legs:	Knotted golden pheasant
Wing:	Turkey tail feather
Overwing:	Deer body hair
Head:	Spun and clipped deer hair

Dave's Cricket

Hook: #8-12, 2XL dry fly
Tail: Black hackle fibers beneath a loop of black yarn
Body: Black yarn palmered with clipped black hackle
Legs: Clipped and knotted black hackle stems
Wing: Black turkey tail feather
Overwing: Deer body hair dyed black
Head: Spun and clipped black deer hair

Dette Beetle

Hook: #12-20 standard dry fly
Thread: Black or white 8/0 Uni
Shell: Black or white deer hair
Body: Peacock herl
Legs: Trimmed stubs of deer hair

Hoover's Periodical Cicada

Hook: #4-6, 2XL dry fly
Body: Orange or green beaver dubbing
Overbody and Head: Black sheet foam, trimmed to shape
Wing: Orange or green Krystal Flash
Legs: Orange or green Sili Legs

Muddled Head Moth

Hook: #6-12, 2XL dry fly
Thread: Tan 8/0 Uni
Body: Tan beaver dubbing
Wing: Deer or elk hair

Green Weenie (Inchworm)

Hook: #10-16, 2XL- nymph
Bead (optional): Black, gold, or copper
Tail (optional): Looped chartreuse chenille
Body: Chartreuse chenille

Afterword

We live in a blessed age for fly fishing. Imagine trying to match North American hatches in the 1700s and 1800s when you had to decipher everything for yourself. There were no true American works on entomology until Louis Rhead's *American Trout-Stream Insects* appeared in 1916. Mr. Rhead's book was the first fly-fishing volume written about the New World's aquatic insects, and much of his stream study occurred in New York's Catskill Mountains.

Mr. Rhead was more entrepreneur than scientist: He ignored all preexisting scientific work and chose his own fanciful names for the insects he encountered. His names were colorful and still bring a smile when I read them. Needle-tail, Brown Buzz, Soldier Drake, and Redbug are all flies Mr. Rhead documented in his book, which includes color plates that allow me to guess at the flies he's talking about, although there's no way to be sure.

Though Mr. Rhead developed a line of fishing flies that were sold exclusively through Leonard Mills and Sons, he didn't include pattern recipes and he didn't tell anglers where they could purchase the necessary materials to construct his flies. He wanted anglers to buy his flies, not tie their own. In some ways this makes Louis Rhead the first modern fly tier. Today, imaginative fly tiers sell the rights to their flies to large companies like Umpqua, Orvis, or Montana Fly Company. The companies tie the flies, and the designer receives a royalty check. Thankfully, the secrecy that was common to fly fishers during Louis Rhead's day has now largely evaporated.

Louis Rhead's book is not very useful for modern anglers and not well regarded by angling historians. His desire to keep his readers needing his flies and his information for matching hatches has made his work almost useless for modern angling purposes. But after Mr. Rhead, several angler/entomologists were inspired to write books that would change our sport forever. And every single one of these books is rooted in research conducted on New York trout streams.

After Louis Rhead, Preston Jennings wrote his *Book of Trout Flies* in 1935. Jennings's book was deeply rooted in the scientific study of the day. We can read Jennings today and know what flies he was talking about. His thoughts and fly patterns are still valid, though the Latin names have certainly changed.

Both Rhead's and Jennings's books were hardbound, mid-sized books. They were intended to be read at home, not along a trout stream. The first pocket-sized book was Art Flick's 1947 *Streamside Guide to Naturals and Their Imitations.* Mr. Flick's book looks very similar to the volume you're holding now. It was small and meant to be taken streamside where an insect could be identified and Flick's methods of imitating it could be put into practice in real time. The book even had photos of a very small number of mayflies. Flick lived in the Catskills and much of his research was conducted on his beloved Schoharie Creek.

Until Ernest Schwiebert wrote *Matching the Hatch,* another pocket-size book, in 1955, little attention was dedicated to waters west of the fabled eastern streams. But though Schwiebert's book was meant to encompass the entire country, some of his research was completed in Pennsylvania's Pocono Mountains and New York's Catskills. Schwiebert's book used color plates of drawings to show the insects he was discussing.

After a nearly 30-year lull (a true testament to the greatness of Schwiebert's book), Al Caucci and Bob Nastasi wrote the first pocketguide that included actual photos of most of the important mayflies, not just the very few that were included in Flick's Streamside Guide. Their *Instant Mayfly Identification Guide* was first published in 1984. It changed the game. No longer were anglers limited by an artist's interpretation or a couple of pictures, now they could take this little pocketguide streamside and compare live insects to actual photographs. Caucci and Nastasi brought the pocketguide to the modern-day angler in a way that no one had done before, and they opened the floodgates for new works

Legendary angling writers like Carl Richards, Dick Pobst, Orvis's Tom Rosenbauer, Thomas Ames, Dave Hughes, and others have all written pocket-sized angling/entomology books that have demystified hatches and angling techniques like never before. Each one of these books has added a new view of trout stream insects and more information for anglers. Each one is worth adding to your library. And it continues: Al Caucci released another pocketguide in 2011 that is perhaps the most comprehensive ever written solely for U.S. mayflies.

The little pocketguide you are reading now could never have been written without the amazing efforts of the men I've mentioned and their books. This book utilizes modern digital cameras that allow me to snap hundreds of photos and choose the best ones for inclusion. I could not imagine doing what Caucci and Nastasi did in their first pocketguide, where they had to collect specimens, shoot them with film, and then wait for the photos to be processed to see if their efforts were even useable. If the pictures weren't useable, new specimens would have to be collected and the process would begin all over again, next year.

I would never claim that this little volume belongs in the amazing, New York pocketguide lineage that I've described. But my book is a little different from the others, and I hope it helped you gain a greater appreciation of the natural world and a deeper understanding of why trout eat, or do not eat, our flies. One thing is certain, without the foundation set forth in the aforementioned books, I could never have written this book.

Bibliography

Though most of the information in this book has come from my own extensive study of aquatic and terrestrial insects throughout New York and Pennsylvania, the following books and websites were a great help to the project by providing additional insights and confirming and advancing my own understanding of the species I have included.

Ames Jr., Thomas. 2009. *Caddisflies: A Guide to Eastern Species for Anglers and Other Naturalists*. Mechanicsburg, PA: Stackpole Books.

Caucci, Al. 2011. *The Mayfly Guide*. New York, NY: Scott & Nix, Inc.

Caucci, Al, and Bob Nastasi. 2004. *Hatches II*. Guilford, CT: Lyons Press.

Funk, David H., Bernard Sweeney, and John K. Jackson. Stroud Water Research Center. 2008. "A Taxonomic Reassessment of the *Drunella lata* (Morgan) Species Complex (Ephemeroptera: Ephemerellidea) in Northeastern North America." (http://www.bioone.org/doi/abs/10.1899/08-026.1) The Society for Freshwater Science. *Journal of the North American Benthological Society*. BioOne. Washington, DC.

Hoover, Gregory A. 2011. "Periodical Cicada" (http://ento.psu.edu/extension/factsheets/periodical-cicada) Penn State University. College of Agricultural Sciences. Entomology, State College, PA.

LaFontaine, Gary. 1981. *Caddisflies*. Guilford, CT: Lyons Press.

Leiser, Eric, and Robert H. Boyle. 1982. *Stoneflies for the Angler*. New York: Knopf.

McCafferty, W. Patrick. 1998. *Aquatic Entomology*. Sudbury, MA: Jones and Bartlett.

——. 2012. "Mayfly Central" (www.entm.purdue.edu/mayfly/index.php). Purdue University, West Lafayette, IN.

Meck, Charles, and Paul Weamer. 2009. *Pocketguide to Pennsylvania Hatches*. New Cumberland, PA: Headwater Books. Mechanicsburg, PA: Stackpole Books.

Mozgai, Dan. 2012. "Cicada Mania" (http://cicadamania.com/cicadas/).

Neuswanger, Jason. 2012. "Troutnut: Aquatic Insect Encyclopedia" (www.troutnut.com).

Rohrbeck, Roger. 2012. "Flyfishing Entomology" (www.flyfishingentomology.com).

Swisher, Doug, and Carl Richards. 2000. *Selective Trout*. New York, NY: Lyons Press.

Weamer, Paul. 2011. *Flyfishing Guide to the Upper Delaware River*. Mechanicsburg, PA: Stackpole Books.

Index